AF552632

PRACTICAL UNDERSTANDING OF CAPABILITY

by implementing *Statistical Process Control*

JAMES C. ABBOTT, P.E.

SECOND REVISED EDITION

Robert
Houston
Smith
Publishers

Practical Understanding of Capability by implementing Statistical Process Control
Second Revised Edition, 1996
by James C. Abbott, P.E.

Abbott Associates, Inc.
150 Executive Center Drive, B37
Greenville, SC 29615
Phone (864) 297-9598
Fax (864) 297-8624
Internet J.C.A@worldnet.att.net

Editor - Gay McLeod
Production Manager - Kelly Boland
Cover Design - Graphic Workshop
Content Advisor - Ron Parker.
Typesetting and Graphic Designer - Krista Spivey.

Published by
Robert Houston Smith Publishers
P. O. Box 1306
Flat Rock, North Carolina 28731-1306
Phone (888) RH-SMITH
Fax (888) RHS-FAX1
Internet rhsp@MAIL.SERVE.COM

Robert Houston Smith Publishers

Library of Congress Catalog Card Number: 95-69102
ISBN 1-887355-02-2

A Note To Those In The "Service" Industry:

All companies produce products. A better distinction than "Product versus Service" is: products with *Tangible* (material) characteristics versus products with *Intangible* characteristics.

Further, all companies produce products which have a mixture of tangible and intangible attributes. Notice, for example, how many commodity items are sold at *convenience* stores and *service* stations.

The concepts in this book apply to any business with any combination of tangible and intangible product characteristics. The difference is: the tangible is typically analyzed with variable data techniques, and intangibles are typically attribute data.

In service industries, the "machine" referred to frequently in this book is usually called: "The Process".

Table Of Contents

Preface

IMPROVEMENT is the watchword of the day, but for improvement to occur, a clear understanding of our product and process capability is essential. The tool to study and assess the capability of a continuous process is Statistical Process Control. This book will show all the subtleties, pitfalls, and benefits associated with using SPC in capability studies.

The Goal

The goal of this book is to provide usable capability knowledge to continually improve while at the same time making our customers happy. What is acceptable and terrific today will be considered terrible tomorrow. I refer to "making the customer happy" as *quality*. Many people use a one-dimensional definition of quality which leads to a trouble zone. For example, by focusing on performance and totally ignoring cost, we risk escalating our cost to an unreasonable level as we try to solve performance problems by simply throwing money at them. To help us avoid this kind of trouble zone, we will develop a definition of quality to use throughout this book as a means of describing our objective for product and process improvement.

One of the biggest problems facing the statistics, analysis, and quality professions is a poor understanding of terms, words, phrases, and concepts. One major purpose of this text is to clarify all the terminology. We will stay technically accurate throughout the book. Let's start this discussion of terms by clarifying the word *quality*.

What is Quality? *Quality* can have many definitions. In *Webster's New World Dictionary*, *quality* is defined as *Any of the features that make something what it is; its characteristic elements; attribute.*

Quality must be put into a perspective for use by all associates. Then quality can become an attitude that will flow throughout an organization in a positive way. A positive approach to quality must be used to get the desired results. Let's see how the definition of quality needs to be expressed.

What Do You Mean By Quality? Write your definition of *quality* below. This will give us a tool to compare your individual impression with my own definition of *quality*.

____

____

____

Our definition of quality is much broader in scope and more precise than the classic quality definition which states that we simply exceed our customer's expectations. I will show how our broadened definition will keep us out of trouble.

In this book I use the following questions to guide us in our definition of quality:

✓ Who Is Our Customer?
✓ What Is the Customer's Intended Purpose?
✓ What Are the Customer's Alternatives?

Quality means more than simply meeting the customer's stated requirements. It means meeting or exceeding the customer's implied or perceived expectations as well. Adding one more dimension, our objective of quality expands to understanding the customer's use of our product. Therefore, the first dimension of quality is exceeding the customer's perceived expectation of our product.

To summarize:
✓ ***Quality is meeting or exceeding the customer's perceived expectations and requirements.***

Our job of quality is not complete yet. We must also reduce our cost. For our product to remain competitive, we must assure that the cost of our product is continually dropping. This gives us many options for directing our monetary savings. We could increase profits, reduce price, increase market share, and invest in the future, to name a few. The second dimension of quality is the continuous reduction of our cost.

Adding this next dimension:
✓ Quality is meeting or exceeding the customer's perceived expectations and requirements ***while reducing cost***.

Still more must be delivered to truly make the customer happy. We must deliver the product to the customer when he wants it. Having a product at a competitive price but which we cannot deliver is disastrous to our quality. Not having any product is just as bad as having a defective product. The third dimension of quality is being able to supply our product when the customer wants it.

Finally our summary definition is:
✓ Quality is meeting or exceeding the customer's perceived expectations and requirements while reducing cost and ***providing the product to the customer WHEN he wants it.***

The three components we've discussed - **Performance**, **Cost**, and **Time** - define quality. This gives us a total definition of quality. Our principle concern is to always strive to provide the best value to the customer. The key word to describe the three dimensions of quality is **VALUE.** A product that is perceived as quality or value by one customer may not be by another. This variation from customer to customer is caused by different uses of the same product.

Many have the mistaken impression that production and quality are conflicting objectives. Nothing could be further from the truth. For us to have production, we must have quality. For us to have quality, we must have production and quantity. To achieve production and quality, we must meet the three quality objectives:

performance (specifications and requirements)
time (schedule)
cost (budget and production).

With these three criteria, we have a clear goal. We must now marshal our efforts to achieve the target goal.

Many times I hear this question: *"Do you want quantity or quality?"*

> **Production** The act or process of producing. The creation of utility. The making of goods available for human wants.

Production is often mistakenly defined as activity, the keeping of machines running, movement of goods from area to area, or the number of goods produced. Goods sent to the warehouse are not counted as production. The only correct time to count production is when we ship the goods the customer wants, and he pays us. The only answer to the question above, "Do you want quality or quantity?" is that we want both **quantity** and **quality**.

Clearly, our definition of quality is extremely broad and encompassing. According to that definition, quality is comprised of three components: performance, cost, and time. Our research, analysis, and decisions must support our goal of delighting our customer (Performance), while reducing our cost (Cost), and providing our product to our customers the moment they want it (Time). We must view this goal as a never-ending journey that continues to improve all three components because no one component is more important than the others. All three must be viewed with equal importance and balance. This balance requires the Business Genius of the Manager of the Future.

Control

The word control is often misused and confused, so clarifying this term is critical to understanding our objectives.

> **Control** a mechanism used to regulate or guide the operation of a machine, apparatus, or system.

The words *consistency* and *predictability* are synonyms for the word *control*. *Consistent* means "repeatable and about the same." *Predictable* means "expected; able to be foretold." Keep these words close at hand for use in explaining the word *control* because when we can repeat and know what to expect, we can run (control) a machine better.

• Consistency • Predictability

The traditional quality assurance department's role is to guarantee that only good product is received by the customer. The purpose of quality assurance is to ensure that the manufactured lots containing excessive defective items are not shipped to the customer.

Quality control has a much broader role in avoiding a deterioration in the product and also continually searching for improvements to the product. Quality control's role thus is a preventive effort.

Quality professionals have taken the lead in pushing the use of the analysis tool called Statistical Process Control. The lead is positive and good, but SPC has received an image of being only the quality control department's tool. Actually though, SPC was designed for use by both operations and management.

A clear understanding of the principles of process management is mandatory to everything we will do. The First Principle of Process Management plays a crucial role in product improvement.

First Principle of Process Management

A fundamental understanding of BOTH the product and process is essential to improvement. Both the product and the process must be described and understood individually and separately. The underlying component for improving the product is the process.

There are two broad areas in which we may collect data. These areas are **product** and **process**. In general, it helps to have data from each area.

A product is...

- **Noun**
- **Result**

When we are discussing products, we are focusing on the *result* of our activities. The **product** is always a noun.

A process is...

- **Verb**
- **Activity**
- **Something that builds the product**

Process discussions are focused on the *activities* that build the product. The **process** is always a verb.

Process

Importance of Process Data. When we monitor the product only, we are aware only of its deterioration or improvement. This awareness of the change in the product does not give us a *reason* for the change. Thus product monitoring is important, but it is:

- Reactionary
- Tolerates waste with no understanding of why
- Accepts improvement with no understanding of why

In my book ***Become the Manager of the Future***, I told the story of my mother learning the art of biscuit-making from my grandmother. Since Grandma was an accomplished biscuit chef but my mother was not, the knowledge transfer was very difficult. Without the proper analysis tools used in the correct areas, Mama burned many biscuits and the baking knowledge was never transferred. My story showed how improper focus keeps us from making improvement. Let's give a brief summary and recap of the First Principle of Process Management.

Without a knowledge of "why," we cannot improve. For us to make Grandma's biscuits, we must understand why things occur the way they do. Product information gives no insight into why we made some good biscuits or why we burnt other biscuits. Since we don't know why we are making good biscuits, we cannot repeat the activities that produced them. **This loss of knowledge is tragic.**

When we monitor the process *and* the product, the process parameters allow us to know the conditions that caused the change. The monitoring of both product and process is essential to:

- Proactive decisions
- Improvements
- Elimination of waste
- Understanding the "why's" of change

The knowledge of "why" allows us to improve continuously. By adding knowledge about the activities required to make a biscuit (process or oven), we begin to learn what is required to make good biscuits. The additional information about the different processes that were used for good biscuits and defective biscuits allows us to continue to repeat the improvements and to discontinue the defects.

In the perfect world of my grandmother and her biscuits, I never saw her check the biscuits when she removed them from the oven. Instead, she kept her focus on the process. Since she knew all the subtleties of baking, she would take corrective action as the oven was cooking the biscuits so that, in a proactive manner, she avoided ever making any burnt biscuits. She was a perfect chef; but unfortunately, most of us are *not* perfect chefs of our processes.

Since we live in a not-so-perfect world, and our knowledge of our process is not nearly so complete as my grandmother's was about her oven, we must continue to monitor both our biscuits and our oven with equal intensity. More precisely, we must monitor both process and product.

Our goal is to use SPC to make economical decisions about actions on the process in order to improve the product.

Balance Risks

All decisions have risk. SPC provides a means of reducing the unknown so as to minimize that risk.

Identifying Good Intentions on the Road to Quality Hell

All stories and examples that are discussed in this book are true. Only the names, products, times, and places have been changed to protect the innocent. As we cover the material, I will show you that many of our well-intentioned efforts wind up causing exactly the reverse of the result we want. Often they not only stand in the way of improvement but also cause our efforts to deteriorate. These are like lead bricks covered with gold paint: the golden bricks of good intentions that we <u>feel</u> are going to help but actually do the reverse.

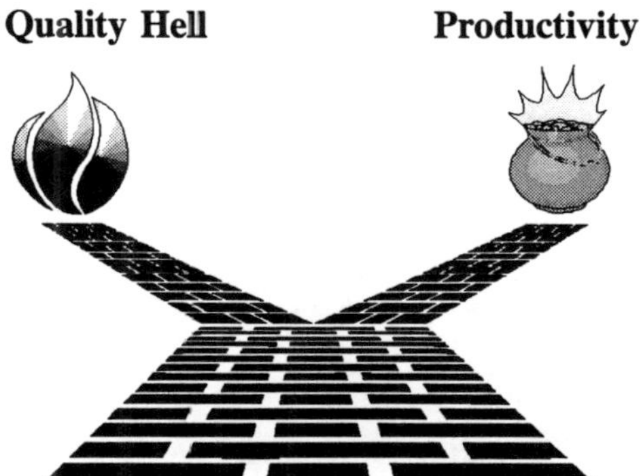

In this book, we will discuss the idea of *Good Intentions*. The following lines from country star Randy Travis's country song "Good Intentions" are very appropriate for our discussions: ***"And I hear tell the road to hell is paved with good intentions, But Mama my intentions were the best."*** The golden bricks that I will describe in this book are paving a road that leads to a horrible, fiery place: a place called Quality Hell.

Many times throughout this book, we will be at a crossroad. One path at this crossroads is paved with the golden bricks of good intentions leading to Quality Hell. The other road is paved with the bricks of Process Management Principles and leads to productivity and quality improvement. We must assure that we have a clear picture of the correct road so that we can plot a course to productivity and improvement.

Good intentions that put bricks on the road to Quality Hell (flawed logic and questionable input to decisions) will be indicated with the quality hell icon to the left. We will also provide practical examples of these good intentions, along with some good common sense advice to avoid these mistakes. Be sure to pay special attention to the gold bricks of good intentions that we will be pointing out.

Continuous Improvement

Improvement requires a clear understanding of the quality objective. All associates must understand that our objective is to continually strive for improvement. This improvement demands that three functions are in place. The first is the philosophy of quality and improvement. Secondly, an objective assessment system must be in place to assure that our course is always directed toward improvement. Finally, execution is key to assure that we actually implement changes and achieve improvement.

Each one of these must be expanded and clearly understood. This book focuses on the objective assessment of a continuous process through Statistical Process Control (SPC). SPC must be in context of an overall improvement model to function as it was intended. Let's now discuss the improvement model.

Improvement Model: Many techniques have been used over the years, with varying degrees of success. I will explain how to put them together to *generate improvement*.

<u>TQM</u>: What we call Total Quality Management (TQM) is called many other things by many other people. One name is "continuous improvement," which is defined as an ongoing effort at improvement. The Japanese refer to this effort as "Kaizen." For us to improve, we must have change, but all change is not improvement. Sometimes we change and unintentionally cause deterioration. The focus of continuous improvement is continuous **change**.

TQM
Kaizen
↓
Continuous Change

<u>SOP</u>: Standard Operating Procedures (SOP) is a management style that forces all procedures and activities to be the same for all machines. No matter what the machines' differences, the SOPs are the same across all machines. This allows management to have complete control of all activities.

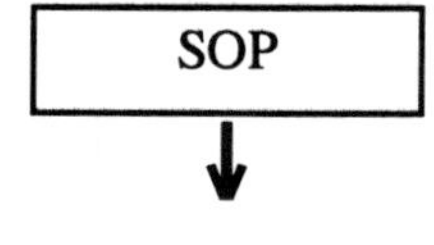

No Change

SOPs have functioned as management's security blanket to assure that we are running the facility based on a set, standard procedure. Since all machines are somewhat different and the SOPs have remained the same, all machines are forced to run **incorrectly.** SOPs discourage change and creativity in order to achieve management simplicity. SOPs are designed to avoid deviation (change) and force the status quo. By following this SOP style, all machines are run inefficiently and not at their optimum level.

<u>Conflict of TQM and SOP</u>: TQM encourages change. SOPs detest change. TQM encourages the very thing that SOPs detest. Therefore, TQM and SOPs are in direct conflict.

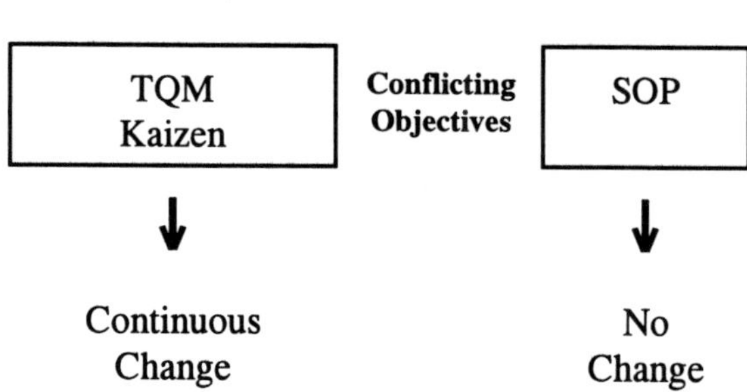

<u>Conflict Resolution</u>: We must resolve the conflict between SOPs and TQM in an

objective manner. What is required is an objective means of assessing all change. These assessments must be immediate and real-time. Changes that result in deterioration are stopped. Changes that bring improvements are incorporated into a revised SOP for a machine. The technique of systematically assessing and incorporating change is called Statistical Quality Control (SQC).

We all know that because people have individual personalities and differences, each person must be treated with his unique traits in mind in order for him to achieve the most he can be. Interestingly, machines are somewhat like people because each machine is different. Even machines that are produced by the same vendor on the same day will have differences. Because we do not expect all machines to be identical (just as all people are not the same), we need to have a distinct SOP for each individual machine and process. SOPs must be targeted to each individual machine.

Improvement Model

The illustration above shows how a TQM or continuous improvement model must be constructed. For TQM to perform correctly, we must incorporate the idea of Standard Operating Procedures (SOP). The classic SOP is a static tool that is seldom used and infrequently updated. To be effective, SOPs must be used on the shop floor and continually updated to reflect our changing world. In Chapter Two, we will explain the structure required to make the improvement model work.

Improvement Model Summary: When we seek to improve our business, we are presented two basic approaches:

Total Quality Management (TQM), often referred to as *continuous improvement*, or *KAIZEN*, is always striving for change.

Standard Operating Procedure (SOP) in the classic sense is always attempting to never have change.

As we discussed earlier, the approaches of **TQM** and **SOP** are always in conflict. The improvement model component called Statistical Quality Control strives to resolve this conflict. Use of this new model allows us to dramatically accelerate our improvement process. We will discuss this model as we progress through this book.

What is Statistical Quality Control? We need to understand that Statistical Quality Control (SQC) is the major umbrella that all the statistical tools fall under.

A tool box that we use in our household holds a variety of tools. A hammer is intended to put nails in a wall. A saw is intended to cut wood. A wrench will tighten and loosen nuts from bolts. We could use a hammer to attempt to remove a bolt, but how effective would this tool (the hammer) be at this job?

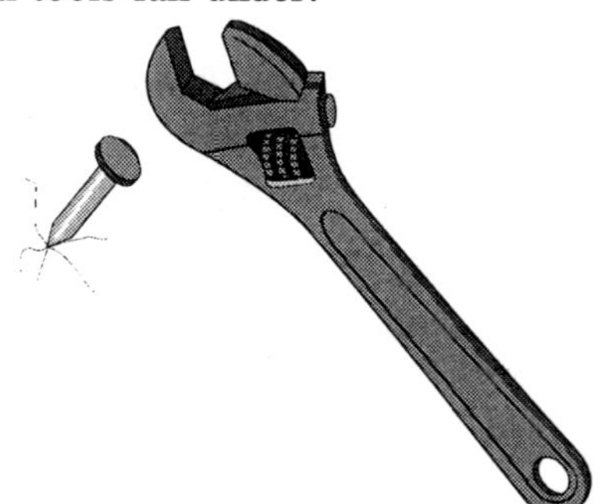

You can easily see how foolish it is to use the wrong tool on the wrong job. Each tool is intended for a particular function and a particular set of problems. As we understand the problems, we will learn which tools are appropriate and how each works. Then we will see how to apply the appropriate tool to each problem. The chart below shows the many tools found in the SQC tool box.

<u>Objective Assessment</u>

Statistical Quality Control

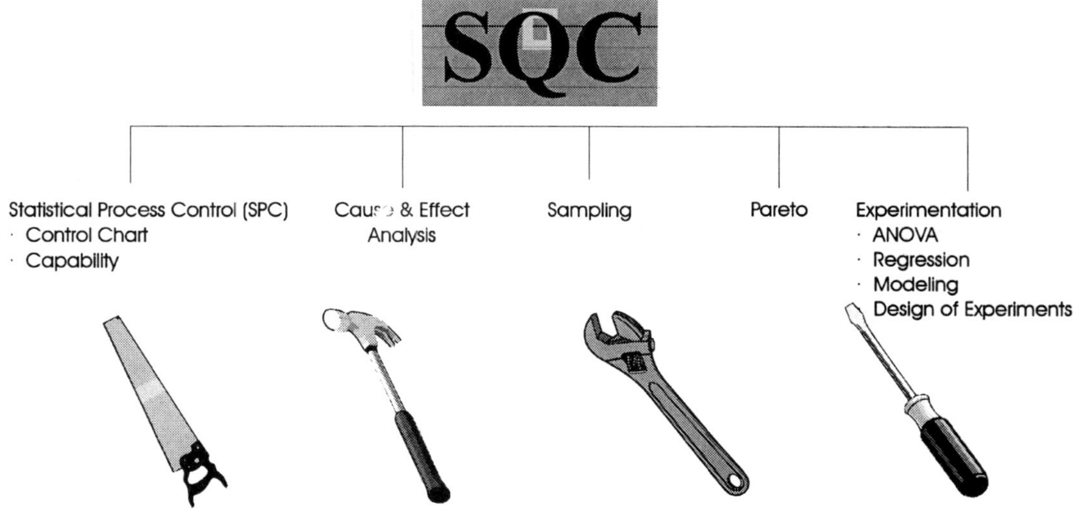

SQC as the Assessment tool Many books on preparing information, reports, or statistics focus all their attention on the analytical tool. Our purpose here is to reverse the classic approach and focus on the situation we will encounter. For each situation, we will examine the specific tools that can be used. Once the appropriate tool has been identified for each situation, we can begin to learn about the tool. Each analytical tool was devised for a particular situation or problem. The following is a quick overview of those situations.

Monitoring of a Continuous Process
- Machines
- Product or Process Consistency
- Noting of Changes
- Assessing Capability

Snapshots of a Moment of Knowledge
- Receipt of Shipment
- Shipping of Product
- Determining Appropriate Test or Sample Sizes
- Polling
- Surveys

Comparative Group Analysis
- Groups
- Machines or Machine Type
- Department or Division
- Laboratories
- Vendors
- Management Reports

Process and Product Optimization
- R & D
- Ongoing Facility
- Minimum Tests
- Parameter Optimum Settings

Analysis of Continuous Data
- Process Parameter Optimization
- Relationships

Focusing Your Staff's Time and Tasks
- Task Dependency
- Critical Task
- Task Priorities
- Critical Resources

This book will focus on **monitoring of a continuous process** through the use of Statistical Process Control. The next section begins our effort to select the correct analytical tool. We will discuss the situation, describe the analytical tool for the situation, and suggest study areas. This book's objective is to explain the SQC tool called SPC (control charts and capability studies). The Statistical Process Control tools will focus on the quality dimensions of performance and cost.

Looking back at the many situations from above, SPC is a razor sharp tool to monitor:

A Continuous Process - Machines, systems, operations
- Product or Process Consistency
- Detection and Notation of Changes
- Assessment of Capability

Understanding Capability by Implementing SPC

Statistical Process Control is a Specialized Tool
Just as a saw from our household toolbox has specific uses, the SPC components of control charts and capability studies are the objective assessment tool for a continuous operation. The control chart detects any changes in the continuous process. Once a change is detected, the capability study is used to assess the impact to the customer.

We must make sure SPC is used for the right situations. Using a great tool for the wrong job can lead to terrible results. If a saw is used to put a nail in the wall, the saw will become dull and the nail will not be solid in the wall. SPC, in like manner, must be used in the correct situation to get proper results.

Since change can be both an improvement and a deterioration, capability studies and control charts must work as a team. The details and subtleties of SPC and the relation of control charts and capability studies to one another will be explained later.

Improvement Model

The first chapter of this book will cover basic statistics. The third chapter covers proper grouping of our data to arrive at good information. The fourth chapter explains the general SPC components of control charts and capability studies. Chapter five through nine explains the construction, use, and interpretation of control charts while chapter Ten explains capability studies.

Let's start our journey into SPC by now learning basic statistics.

Chapter One: Basic Statistics

What is STATISTICS? The dictionary states that statistics are *1) Facts or data of a numerical kind, assembled, classified, and tabulated so as to present significant information about a given subject.* 2) *the science of assembling, classifying, tabulating, and analyzing such facts or data.*

Researchers, analysts, and anyone studying numbers require a common communication language. Statistics is that common communication language.

Statistics is comprised of a set of tools used to prepare meaningful information. The purpose of this chapter is to present some statistical tools of practical value. Our first look at statistics will help us to understand the tools called **point estimates**. Statistics is the informational eyes of the decision-maker. Since we must make decisions in the face of uncertainty, statistics offers the tools to assist in these situations.

Our objective must always be to provide as clear a picture of the data as possible while also keeping our volume of numbers as low as possible. To assure that we never present improper information, we must identify the proper statistical procedures and the areas that will cause our information to be flawed.

Report and Analysis Flaws

Objective of Analysis = Knowledge Base The analysis of our data provides the cohesion to build the strategies and tactics that will drive our improvement. It must always be done with clear guidelines that will avoid any tainted results. The product of our analysis is a knowledge base.

It is not enough to simply trace the history of events. Our analysis also must uncover the *causes* of the effects. Since an effect is not caused by simply one item, our analysis must continuously strive to uncover more and more of the causes and how much of the effect can be assigned to one cause.

The analysis must endeavor to uncover the who, what, where, and when of the cause that is driving the effect. This is not a one-time effort but a continuing and on-going search to increase our Knowledge Base. We must clearly understand and adhere to the science of how to analyze the who, what, where, and when so that we avoid any improper or inappropriate information that might drive us to make improper decisions.

Management Reporting Flaws. Our objective is to assure that we understand the potential flaws of improperly prepared data. In order to avoid these costly mistakes, we must have a clear set of guidelines so that we can prepare information that allows us to make accurate and intelligent decisions.

Understanding Capability by Implementing SPC

Flaw 1 - Technique We must understand and use proper techniques while we avoid the mistakes that can be caused by using improper techniques.

Technique The systematic procedure by which a complex or scientific task is accomplished.

Statistics brings to us the consistent and common language of correctly preparing our numbers with proper technique.

Flaw 2 - Grouping Grouping can lead to a different set of problems. These problems must be clearly understood so that all of our studies will use the proper groups.

Grouping 1. A number of individuals or things considered together because of similarities. 2. A class or collection of related objects or entities

The efforts involved in the assurance that we are properly grouping is an integral part of statistical analysis.

Reports are used often to show when things are bad. Being a child of the 50s, I grew up watching TV. One of my favorite programs was DRAGNET. The main character was Joe Friday played by Jack Webb. His watch word and the theme of the program was always "Just The Facts Ma'am." We need to pattern ourselves after Joe Friday, neither looking for Good nor Bad but always looking for the Facts.

This chapter will explain proper statistical technique while Chapter Three will clarify the need for and types of grouping required for proper information to be presented.

Our first analytical study will revolve around a common problem. We will analyze the electricity usage for my household. The table below shows the household kilowatt consumption of my house for the years 1980, 1981, 1982, and the partial year for 1983.

For statistics to work effectively, a clear objective must be stated. Therefore, the information gained from this study will be used to plan a household budget so that enough funds may be set aside to pay for the electricity consumed.

These values represent the kilowatt consumption for our house.

ELECTRICITY USAGE

	1980	**1981**	**1982**	**1983**
JANUARY	2749	4081	4982	3936
FEBRUARY	5405	5004	4299	5024
MARCH	3936	2708	3793	3239
APRIL	2232	2423	1870	
MAY	1079	1493	1514	
JUNE	1330	2197	1713	
JULY	2616	2918	1890	
AUGUST	2820	2041	1897	
SEPTEMBER	2321	2041	1923	
OCTOBER	1583	1494	1316	
NOVEMBER	1877	1859	1843	
DECEMBER	3439	3936	2919	

Formulas In mathematics, a clear way to communicate our calculations is required, and this communication is achieved through the use of formula. A formula calculates a *value* based on numbers mathematically computed. In this book, all formulas will contain variables to compute the value. The symbols for the math operators that will be used in this book are listed below:

( )	order of operations
*	multiplication
/	division
+	addition
-	subtraction

Raising to a power means multiplying the number by itself a given number of times.

$$X * X = X^2$$

The symbol ^ is used by software packages to identify the raising to a power of a number. The raising of 2 to the second power (2^2) yields the result of 4 while two to the third power (2^3) is 8.

Taking a root is the reversing of the power-raising process.

$$\sqrt{X^2} = X \qquad \sqrt[3]{X^3} = X$$

The square root of 4 is 2 while the cube root of 8 is also 2.

Summation of a set of numbers can be expressed several ways. The simplest is to write the complete expression. For our example of totaling the 1980 kilowatt consumption, each month - January through December - will be represented using the summation symbol. The use of three dots indicates the repeating of the pattern until the signified value is reached.

Total KWH consumption = $X_{Jan} + X_{Feb} + X_{Mar} + \ldots + X_{Dec}$

The Greek symbol *sigma* below is the symbol used to identify the summation of a series of numbers.

$$\sum_{i=1}^{N} X_i$$

The above equation can be used rather than writing out the complete expression like the first method.

As we progress through the book, the above symbols will be used to build our statistical and mathematical expressions. Many computer spreadsheets have statistical functions. As I discuss each statistical function, I will identify the *Lotus* and *Excel* functions that allow you to expedite the process of analysis.

Spreadsheets (*Lotus, Excel, Quattro*) have built in math functions which are actually built-in formulas. A statistical function calculates its value based on the values in a worksheet range or a group of cells. A *range* is spreadsheet terminology for a rectangular block of cells.

@sum(RANGE) The sum of the values in RANGE in *Lotus*.
=sum(RANGE) The sum of the values in a RANGE for *Excel*.
These functions are equivalent to the algebraic functions below:

$$X_1+X_2+X_3+...X_N$$

where N = the number of observations
OR

$$\sum_{i=1}^{N} X_i$$

where N = the number of observations

```
B6: @SUM(B2..B4)                                                READY
          A         B         C         D         E         F         G         H
1
2                   1
3                   2
4                   3
5          ---------
6                   6
```

TO ENTER A SUM FUNCTION
Other functions also follow this pattern:
STEPS
1. Identify the range to be computed.
2. Position the cell pointer on the cell where the sum is to appear. (OR, begin after a math operator in a formula.)
3. Type the following in your spreadsheet In Lotus *@SUM(* or In Excel *=SUM(*
4. Specify the RANGE:
   1. MOVE the cell pointer to the upper left corner of the range to be computed.
   2. Press period (.) to ANCHOR the cell pointer in Lotus and Colon (:) in Excel.
   3. Move the cell pointer to the opposite corner of the range to HIGHLIGHT the range.
5. Type the closing parenthesis:)
6. Continue the formula, if required. When the formula is complete, press RETURN to enter the formula.

Let's see how to put our math to use in statistics. Proper statistical technique requires the monitoring of the three components: central tendency, variability, and distribution. Each component must be understood and used correctly. Our first tool will monitor the data's central tendency or balance point.

What is Balance? When we were children, we used to play on a See-Saw or Teeter-Totter. One of the main things we tried to do was get the See-Saw to balance two (2) people on opposite ends of a board. The spot the board rests on is called the *fulcrum.*

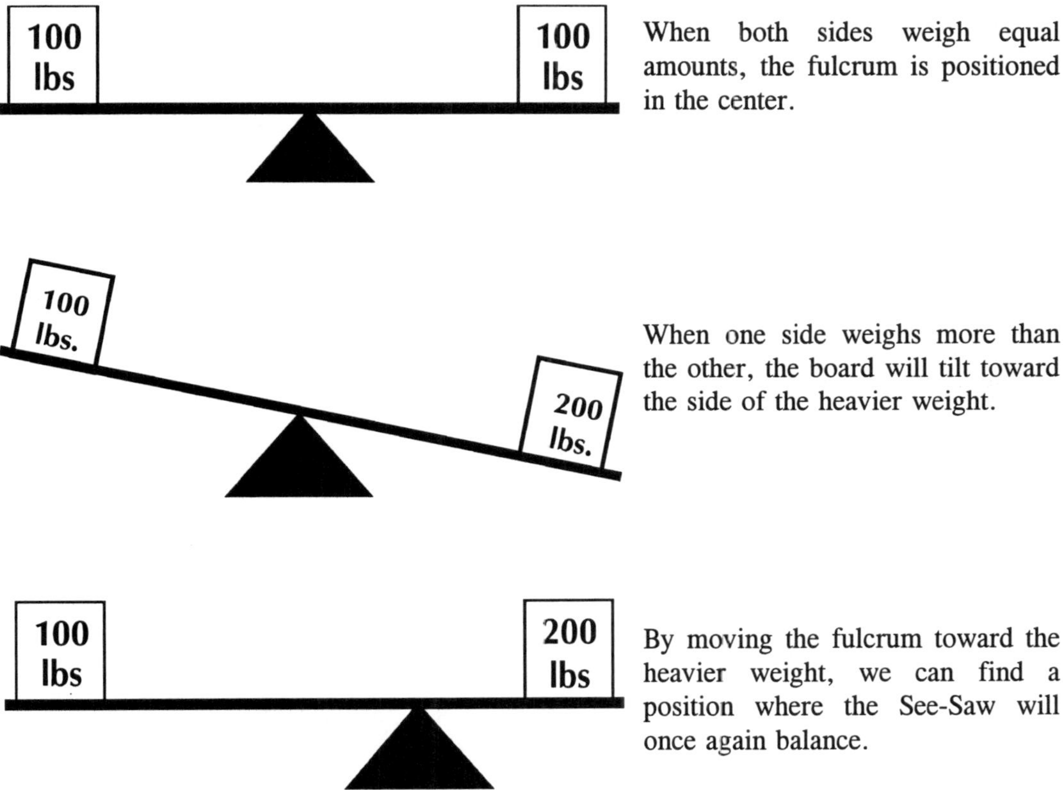

When both sides weigh equal amounts, the fulcrum is positioned in the center.

When one side weighs more than the other, the board will tilt toward the side of the heavier weight.

By moving the fulcrum toward the heavier weight, we can find a position where the See-Saw will once again balance.

This balance point is critical information in many fields like physics and mechanical engineering. Knowledge about balance associated with numeric measurements is essential to statistics. The balance point of our numbers, like the fulcrum, is called **central tendency**.

Central Tendency Since central tendency is critical to understanding our numeric measurements, monitors of it must be established. Our first monitor of central tendency is **average**. Average is a point estimate of the measurement of the central tendency of our data. Average is a statistical tool to monitor the central tendency and is referred to as mean, arithmetic mean, and simply - average. All the different names are describing the same monitor. Average is the balance point of all the data points giving equal weight to each value.

Spreadsheet functions calculate basic descriptive statistics on the values found in worksheet cells. These spreadsheet functions can compute this phenomenon called **central tendency**.

TOTAL @sum(RANGE) Sums the values in RANGE.

$$X_1+X_2+X_3+\ldots X_N$$

Where N = the number of observations

OR

The expression can also be written using our Greek symbols as follows.

$$\sum_{i=1}^{N} X_i$$

AVERAGE Average divides the sum of the measurements by the number of measurements. The average thus monitors the **central tendency**.

The following are the spreadsheet functions that compute average.

@avg(RANGE) is the Lotus spreadsheet function to compute the average of the values.

=average(RANGE) is the Excel spreadsheet function to compute the average of the values.

The algebraic expression for average is shown below.

$$\frac{X_1+X_2+X_3+\ldots X_N}{N}$$

OR

$$\frac{\sum_{i=1}^{N} X_i}{N}$$

The average may also be referred to as the following.

X_{Ave}

OR

$\bar{X}$

For our earlier electricity data, to compute the average kilowatt consumption for the year 1980, individual consumption for each of the twelve months is added and then divided by the total number of months (twelve). The equation would read

Total KWH consumption $= X_{Jan} + X_{Feb} + X_{Mar} + \ldots + X_{Dec}$

$= 2749 + 5405 + 3936 + \ldots + 3439$

$= 31387$

The 1980 total KWH consumed is 31387 which is then divided by 12 to arrive at the average which is 2616 Kilowatt hours.

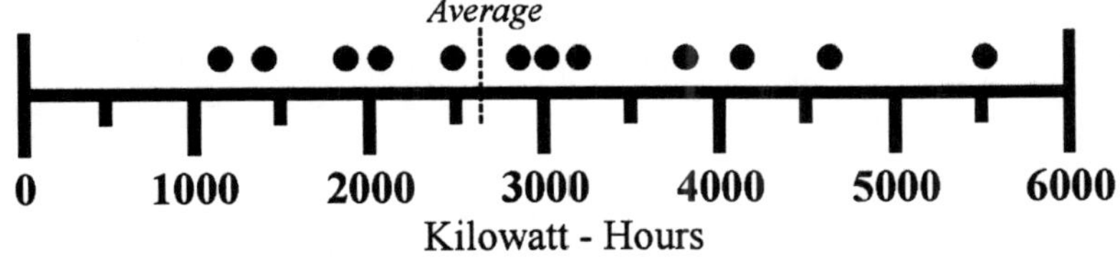

Note how clearly the average shows the center point of our electricity consumption.

Never use Average as a stand alone measure of a set of values!! **Central tendency** should not be discussed without a companion monitor of variability.

Misleading Averages

The three sets of data to the right are intended to demonstrate that simply looking at an average can give an erroneous impression. Suppose we are trying to compare the price point of three motel chains.

Motel Location	A	B	C
Greenville, SC	25	25	10
Charlotte, NC	27	25	15
Atlanta, Ga	24	25	20
Birmingham, AL	25	25	25
Nashville, TN	26	25	20
Indianapolis, IN	26	25	15
louisville, KY	24	25	10
Lexington, KY	23	2	25
Cincinnati, OH	24	25	35
Columbus, OH	26	25	15
Denver, CO	27	25	40
Las Vegas, NV	23	25	35
Dallas, TX	25	25	30
Los Angeles, Ca	25	25	25
San Francisco, CA	24	56	30
Newark, NJ	26	25	35
Toledo, OH	25	25	40
average	$25	$25	$25

Notice that all three of the motel chains wind up with exactly the same average of $25 per night, but all the values of the data set are radically different. The first set of values are relatively close to the average. The second set has two values extremely removed from all the other values. The third set has a widely disbursed set of values. The only common link between values is that they all have an average of $25.00

My family has convinced me that vacations can be a wonderful experience. We have planned and been on several vacations over the last six years. Imagine planning a vacation's sleeping room cost based on the improperly prepared motel statistics from above. The stand alone averages create a false picture that could cause the cost to be way above our expected cost. An otherwise wonderful vacation is destroyed when our vacation money is exhausted. Statistics must provide a clear summary picture of our numbers. Improperly prepared statistical technique must be avoided to assure that these kinds of prediction mistakes are avoided.

We must always get a complete picture of our data. Nothing is wrong with the average - it simply shows only part of the picture.

ELECTRICITY USAGE

	1980	1981	1982	1983
JANUARY	2749	4081	4982	3936
FEBRUARY	5405	5004	4299	5024
MARCH	3936	2708	3793	3239
APRIL	2232	2423	1870	
MAY	1079	1493	1514	
JUNE	1330	2197	1713	
JULY	2616	2918	1890	
AUGUST	2820	2041	1897	
SEPTEMBER	2321	2041	1923	
OCTOBER	1583	1494	1316	
NOVEMBER	1877	1859	1843	
DECEMBER	3439	3936	2919	
Average	2616	2???	????	

Returning to our household kilowatt hour (kwh) consumption, the average is 2616 kwh for the year of 1980. Compute the remaining yearly average and make an interpretation of the results.

Now let's expand on our knowledge and add another statistical component that needs to complement the average. This component is the amount or magnitude of spread associated with our measurements.

Magnitude of Spread

SPREAD

We have seen from the motel example that an average by itself can be very misleading. This would be like saying that all people who weigh the same amount look the same. As our little diagram shows, the two individuals above are very different. One area of their difference is that one person has a spread and the other does not. If we were analyzing people's weights, we would want to make sure that we had an additional technique to monitor this spread. The magnitude of this spread can be crucial to our view of a person.

When we are doing statistical analysis, a technique for monitoring spread is required just as it is with a person's weight. The magnitude of this spread is vital just like it is in monitoring a person. The name for the magnitude of the spread is called **variability**. Next we will begin to discuss a series of techniques for monitoring variability or the magnitude of the spread.

Variability We need to have a way of distinguishing the difference between the three sets of motels from the prior discussion. The means of doing this is called **monitoring the variability**. Our first technique for monitoring the variability of the measurements is by viewing the highest and lowest values of the data.

Spreadsheet functions can calculate the basic descriptive statistics of variability of the values found in worksheet cells. Thus the spreadsheet functions can monitor the amount of spread associated with the data. These spreadsheet variability functions are shown below.

@min(RANGE)	The smallest of the values in *Lotus*.
@max(RANGE)	The largest value in Lotus.
=min(RANGE)	The smallest of the values in *Excel*.
=max(RANGE)	The largest value in Excel.

When using minimum and maximum, two values are always required. This can be reduced by using the distance between the maximum and minimum values. This distance is called the range value. Sometimes the range is abbreviated and called the **R-value**.

Range — The range is a measure of the variability of a set of numbers. The smaller the range, the lower the variability or the amount of spread.

Range = Maximum - Minimum
In Lotus Range equals @MAX() - @MIN()

Range = Maximum - Minimum
In Excel Range equals =MAX() - MIN()

Using the electricity data to compute the kilowatt consumption range for the year 1980, the high value of each month's consumption is 5405 and the low value 1079. The range for 1980 KWH consumption equals X_{Max} - X_{Min}. The values then can be placed in the equation as shown below to compute the range.

$$\text{Range} = X_{Max} - X_{Min}$$

$$= 5405 - 1079$$

$$= 4326$$

Finally, the 1980 range KWH consumed of 4326 Kilowatts is computed. The chart below shows our monthly kilowatt hour consumption on a graph along with the average, maximum, minimum, and range values.

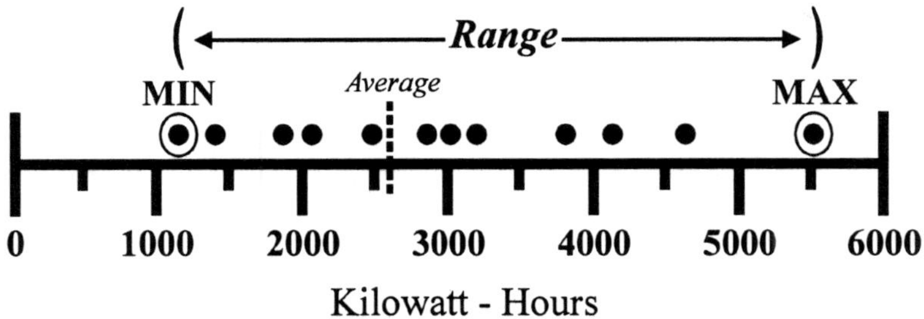

How do I Interpret and Use Range?

As the range increases, the magnitude of the variability (spread) increases. Should the range be zero, all values are the same and the average is a perfect indicator of each data point. The larger the range, the more variability the data has. As the range increases, the average becomes a less precise indicator of the actual values.

Interpreting Range The 3 sets of data below demand some monitor of variability to clearly depict the different hotel chains.

Motel Location	A	B	C
Greenville, SC	25	25	10
Charlotte, NC	27	25	15
Atlanta, GA	24	25	20
Birmingham, AL	25	25	25
Nashville, TN	26	25	20
Indianapolis, IN	26	25	15
Louisville, KY	24	25	10
Lexington, KY	23	2	25
Cincinnati, OH	24	25	35
Columbus, OH	26	25	15
Denver, CO	27	25	40
Las Vegas, NV	23	25	35
Dallas, TX	25	25	30
Los Angeles, CA	25	25	25
San Francisco, CA	24	56	30
Newark, NJ	26	25	35
Toledo, OH	25	25	40
average	$25	$25	$25
Maximum	$27	$56	$40
Minimum	$23	$2	$10
Range	$4	$54	$30

By using range and average together, we can begin to understand the differences in the three sets of data. Chain A has a mean cost of $25 and a range of $4. This drives us to the conclusion that Chain A's rates are all close to the mean. Chain B has a mean cost of $25 and a range of $54. We would conclude that Chain B's rates are very widely spread. Chain C has a mean cost of $25 and a range of $30. We would conclude that Chain C's rates are also widely spread.

Since the range values for both Chain B and Chain C are both high values, we might conclude that both chains charge similar rates. We can see by analyzing the data, however, that the costs of both B and C are very different. Chain B has 2 very extreme values with the remaining rates all being equal to $25. All of Chain C's rates are extremely different. Therefore, the range value fails us in differentiating between these two motel chains.

We must always get a complete picture of our data. Nothing is wrong with the average, but it simply shows only part of the picture. To add more precision to our study, we must monitor both central tendency and variability together. Neither one can be used without the other.

Properly prepared statistical information can be a wonderful tool for planning and predictions for all decision-making including the planning of our vacation.

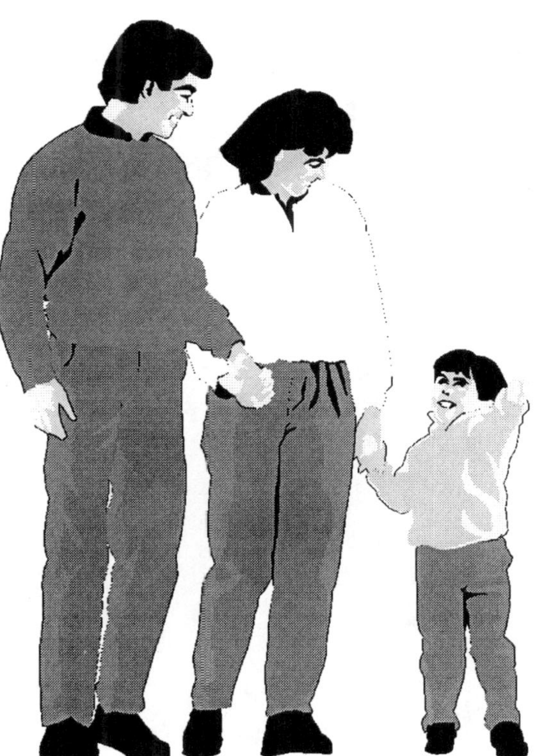

Now our family can anticipate a pleasant vacation because our statistics have been prepared with a more vivid picture which now includes both central tendency and variability. We must always be careful to follow proper statistical technique.

Our plans now can be made with reliable and predictable results. Using this proper statistical technique, fun filled future vacations can be assured while avoiding the disasters of exceeding our planned vacation budget.

The kilowatt average and range now work together to allow us to begin to plan our future electricity needs.

ELECTRICITY USAGE

	1980	1981	1982	1983
JANUARY	2749	4081	4982	3936
FEBRUARY	5405	5004	4299	5024
MARCH	3936	2708	3793	3239
APRIL	2232	2423	1870	
MAY	1079	1493	1514	
JUNE	1330	2197	1713	
JULY	2616	2918	1890	
AUGUST	2820	2041	1897	
SEPTEMBER	2321	2041	1923	
OCTOBER	1583	1494	1316	
NOVEMBER	1877	1859	1843	
DECEMBER	3439	3936	2919	
Average	2,616	2,683	2,497	4,066
Range	4,326	3,511	3,666	1,785

The average for 1980 is 2616 KWH with a range of 4326 KWH. A range of this magnitude in relation to its average is quite large. Our conclusion would indicate that large swings from month to month should be anticipated if 1980 is representative of the future.

The year 1981 also has a large range of 3511 KWH in relation to its average of 2683 KWH which would reinforce our conclusion from the analysis of 1980.

Similarly, 1982 has a large range of 3666 KWH in relation to the average of 2497.

Since 1983's KWH consumption is incomplete, no conclusion should be made from this year's data.

To increase our information about the spread of our data, a more advanced method of monitoring variability is required. This improved method should include all our data.

Average Range

As an alternative to a simple range, we could possibly develop a range value for each data point. This could be accomplished by comparing each data point to the monitor of central tendency called **average**.

Range = Data Point - Average
= $X_i - X_{AVE}$

ELECTRICITY USAGE		
	1980	**Range**
JANUARY	2749	133
FEBRUARY	5405	2789
MARCH	3936	1320
APRIL	2232	-384
MAY	1079	-1537
JUNE	1330	-1286
JULY	2616	0
AUGUST	2820	204
SEPTEMBER	2321	-295
OCTOBER	1583	-1033
NOVEMBER	1877	-739
DECEMBER	3439	823
SUM	31387	
AVG	2616	
Max	5405	
Min	1079	
Range	4326	

For the electricity used in 1980, we compute the range for January by taking the January value of 2749 KWH and subtracting the average of 2616 KWH. The result yields a positive 133 as in the table to the right.

The February range requires subtracting February's value of 5405 KWH by the average of 2616 KWH. The February range is a positive 2789 KWH. This process is continued for each month as in the table to the right also.

The problem with a range for every data point is that as we increase our number of tests or samples, the number of range values becomes unmanageable. Computing an average of the ranges becomes an optional solution.

$$R_{AVE} = \frac{\sum_{U=1}^{N} (X_i - X_{AVE})}{N}$$

ELECTRICITY USAGE		
	1980	**Range**
JANUARY	2749	133
FEBRUARY	5405	2789
MARCH	3936	1320
APRIL	2232	-384
MAY	1079	-1537
JUNE	1330	-1286
JULY	2616	0
AUGUST	2820	204
SEPTEMBER	2321	-295
OCTOBER	1583	-1033
NOVEMBER	1877	-739
DECEMBER	3439	823
SUM	31387	-0
AVG	2616	
Max	5405	
Min	1079	
Range	4326	

To solve the problem of an unmanageable number of range values, we could try to calculate an **average range.** This idea is great. It allows us to take into account all the data points and give them equal weight. The problem we encounter is that some of the values are above the mean and some are below the average. This is because the average is a monitor of central tendency. This is exactly what we expected to see! Since the average is equidistant from all the data points, when we compute an average range, the pluses and minuses offset and result in a zero(0).

The resulting zero creates a problem for our use as a monitor of variability. So then, for averaging range, a method which compensates for the minus signs must be found.

Standard Deviation

To remove the sign for each range, we can square each range value. The squaring effort shown in the last column always returns a positive sign.

ELECTRICITY USAGE

	1980	Range	Range Squared
JANUARY	2749	133	17800
FEBRUARY	5405	2789	7780845
MARCH	3936	1320	1743500
APRIL	2232	-384	147136
MAY	1079	-1537	2361088
JUNE	1330	-1286	1652725
JULY	2616	0	0
AUGUST	2820	204	41786
SEPTEMBER	2321	-295	86779
OCTOBER	1583	-1033	1066228
NOVEMBER	1877	-739	545505
DECEMBER	3439	823	678015

$$R_{AVE} = \frac{\sum_{i=1}^{N} (X_i - X_{AVE})^2}{N}$$

The formal name for this value is the **variance**.

ELECTRICITY USAGE

	1980	Range	Range Squared
JANUARY	2749	133	17800
FEBRUARY	5405	2789	7780845
MARCH	3936	1320	1743500
APRIL	2232	-384	147136
MAY	1079	-1537	2361088
JUNE	1330	-1286	1652725
JULY	2616	0	0
AUGUST	2820	204	41786
SEPTEMBER	2321	-295	86779
OCTOBER	1583	-1033	1066228
NOVEMBER	1877	-739	545505
DECEMBER	3439	823	678015
@SUM	31387	Sum of Squares=	16121409
@AVG	2616	Number of Data=	12
@Max	5405	SS/n=	1343451
@Min	1079	Standard Deviation=	1159
Range	4326		
@Std	1159		

To get our unit of measure back to its original state, we take the square root of the above value giving the following equation:

$$\text{Standard Deviation} = \sqrt{\frac{\sum_{i=1}^{N} (X_i - X_{AVE})^2}{N}}$$

The resulting equation is called **standard deviation**. The square root of the variance also yields the standard deviation.

Since we are squaring the range component of this equation, we may reverse the values to give the following formula:

$$\text{Standard Deviation} = \sqrt{\frac{\sum_{i=1}^{N} (X_{AVE} - X_i)^2}{N}}$$

The standard deviation is abbreviated as *s* or *std* while variance uses s^2.

Standard Deviation Interpretation: As the magnitude of the standard deviation increases, the variability increases. Should the standard deviation be zero, all values are the same. Thus average becomes a perfect indicator of all the data with a standard deviation equal to zero. Standard deviations are all positive numbers.

Small Sample Correction of Standard Deviation

The equation below uses sample size (n) as the divisor. Averages of samples are imperfect estimates of population central tendency. Repeated samples produce a range of averages. Standard deviations of samples are also imperfect estimates of the population variability. Repeated samples produce a range of standard deviations. Unlike sample averages, standard deviations of samples tend to underestimate the variation of the population. When the sample size falls below 30, this difference becomes significant. Below 30 samples, the formula provides poor results in tracking the variability of the measurements for the population.

$$\text{Population Standard Deviation} = \sqrt{\frac{\sum_{i=1}^{N} (X_{AVE} - X_i)^2}{N}}$$

The solution to this problem is to subtract one from the sample size or *n*. This tracks the true variability of the population measurements. This function is called the **small sample correction factor**.

$$\textit{Small Sample Standard Devation} = \sqrt{\frac{\sum (\bar{X}-X_i)^2}{N-1}}$$

The only difference in the two standard deviation equations is the minus one (-1) in the divisor. When the sample size is large, the population formula should be used, but the difference is so small that generally the small sample correction formula for standard deviation is used.

Spreadsheet Variability functions

These functions calculate basic descriptive statistics on the values found in a RANGE of worksheet cells. **Standard deviation** is the measure of variability of data points giving equal weight to each data point. Another way of describing standard deviation is to say that it is an average range if average is the base point from which all data is referenced.

Variance

@var Lotus' variance of the population. It does not use the small-sample correction factor.
@vars Lotus' variance of the sample.
=var Excel's variance of the sample.
=varp Excel's variance of the population.

Sample $$s^2=\frac{\sum(\bar{X}-X_i)^2}{N-1}$$

Population $$s^2=\frac{\sum(\bar{X}-X_i)^2}{N}$$

Standard Deviation

@std Lotus' standard deviation for the population
@stds Lotus' standard deviation of the sample
=stdev Excel's sample standard deviation
=stdevp Excel's population standard deviation

Sample $$S=\sqrt{\frac{\sum(\bar{X}-X_i)^2}{N-1}}$$

Population $$S=\sqrt{\frac{\sum(\bar{X}-X_i)^2}{N}}$$

In the statistical community the term signa shown as σ is used to signify the population's variability. In addition the statistical community uses the term mu shown as μ is used to signify the population's central tendency.

Interpreting Standard Deviation

Now our additional monitor of variability, standard deviation, should be used. Remember, this value must be used in conjunction with average.

Motel Location	A	B	C
Greenville, SC	25	25	10
Charlotte, NC	27	25	15
Atlanta, GA	24	25	20
Birmingham, AL	25	25	25
Nashville, TN	26	25	20
Indianapolis, IN	26	25	15
louisville, KY	24	25	10
Lexington, KY	23	2	25
Cincinnati, OH	24	25	35
Columbus, OH	26	25	15
Denver, CO	27	25	40
Las Vegas, NV	23	25	35
Dallas, TX	25	25	30
Los Angeles, CA	25	25	25
San Francisco, CA	24	56	30
Newark, NJ	26	25	35
Toledo, OH	25	25	40
Average	$25	$25	$25
Maximum	$27	$56	$40
Minimum	$23	$2	$10
Range	$4	$54	$30
Standard Deviation	$1	$9	$10

For the three motel chains above, we would draw the following conclusions. Chain A's having an average of $25 and a standard deviation of $1 indicates that the rates in this chain are very tightly clustered. Chain B's having the same average of $25 but a standard deviation of $9 indicates that the rates in this chain are widely spread. Chain C's having an average of $25 and a standard deviation of $10 indicates that the rates in this chain are also very widely spread.

For this example, using monitors of central tendency and variability, Motel A is clearly different from Motels B and C. By viewing the actual rates of Motels B and C, we see that additional techniques may be required for analysis of these motels. We will discuss these in a later chapter.

ELECTRICITY USAGE				
	1980	**1981**	**1982**	**1983**
JANUARY	2749	4081	4982	3936
FEBRUARY	5405	5004	4299	5024
MARCH	3936	2708	3793	3239
APRIL	2232	2423	1870	
MAY	1079	1493	1514	
JUNE	1330	2197	1713	
JULY	2616	2918	1890	
AUGUST	2820	2041	1897	
SEPTEMBER	2321	2041	1923	
OCTOBER	1583	1494	1316	
NOVEMBER	1877	1859	1843	
DECEMBER	3439	3936	2919	
Average	2,616	2,683	2,497	4,066
Range	4,326	3,511	3,666	1,785
Std	1,211	1,112	1,211	900

The kilowatt average, range, and now standard deviation work together to allow us to plan our future electricity needs even better. The standard deviation for the three years of 1980, 1981, and 1982 are very similar: 1211, 1112, and 1211 respectively. This will allow us to conclude that our monitor of variability is a good measure. The magnitude of the standard deviation is still very high in relation to the averages, thus forcing us to plan for large swings in the amount of electricity that will be consumed each month.

Where is the Spread Located

From our prior discussion, Motels B and C both have averages and standard deviations that are relatively similar. A new technique is required to show the difference for examples like this. The problem is that the magnitude of the central tendency and variability are similar. This new technique needs to show a different dimension. This dimension needs to represent not the magnitude of the spread but the location of the spread. Notice the three drawings below of people who weigh the same amount.

The first example (the person to the left) is a person whose shape is that of a body builder. His largest mass is located in the top of his body.

The second example (the person to the left) is a person who weighs the same as the first person. The difference between these two people is that the second person's mass is located in the center of his body.

The third person (located to the left) is a person weighing about the same as the first two individuals. This person's mass is located at the bottom of his body.

All three drawings are similar but all three are different. The technique we need to differentiate between the three people needs to show the location of their spread. The drawings above do an excellent job of presenting this information. Let's discuss how we can prepare a picture of data that will present the location of the spread just like the pictures above present the people's location of their spread.

Now let's begin to analyze a new set of data which we want to convert into significant information. We'll begin with a set of data about the presidents of the United States from the first president to the fortieth president. Our information includes the term number of each president, his first and last names, the state and year of his birth, and his age at his inauguration. We want to prepare a report based on this data.

The particular aspect we will analyze is the age at inauguration of each president. We have already calculated the average age of the presidents at inauguration to be 55.55 years of age. The standard deviation is 6.07 years. In other words, the balance point of the presidents is 55 years of age with the amount of spread being 6 years of age. Let's now begin to analyze the location of the spread or **distribution** of the presidents' ages.

TERM	FIRST NAME	LAST NAME	HOME STATE	BIRTH_YY	INAUGURATION AGE
1	George	Washington	Virginia	1732	57
2	John	Adams	Massachusetts	1735	62
3	Thomas	Jefferson	Virginia	1743	58
4	James	Madison	Virginia	1751	58
5	James	Monroe	Virginia	1758	59
6	John	Adams	Massachusetts	1767	58
7	Andrew	Jackson	South Carolina	1767	62
8	Martin	Van Buren	New York	1782	55
9	William	Harrison	Virginia	1773	68
10	John	Tyler	Virginia	1790	51
11	James	Polk	North Carolina	1795	50
12	Zachary	Taylor	Virginia	1784	65
13	Millard	Fillmore	New York	1800	50
14	Franklin	Pierce	New Hampshire	1804	49
15	James	Buchanan	Pennsylvania	1791	66
16	Abraham	Lincoln	Kentucky	1809	52
17	Andrew	Johnson	North Carolina	1808	57
18	Ulysses	Grant	Ohio	1822	47
19	Rutherford	Hayes	Ohio	1822	55
20	James	Garfield	Ohio	1831	50
21	Chester	Arthur	Vermont	1830	51
22	Grover	Cleveland	New Jersey	1837	48
23	Benjamin	Harrison	Ohio	1833	56
24	Grover	Cleveland	New Jersey	1837	56
25	William	McKinley	Ohio	1843	54
26	Theodore	Roosevelt	New York	1858	43
27	William	Taft	Ohio	1857	52
28	Woodrow	Wilson	Virginia	1856	57
29	Warren	Harding	Ohio	1865	56
30	Calvin	Coolidge	Vermont	1872	51
31	Herbert	Hoover	Iowa	1874	55
32	Franklin	Roosevelt	New York	1882	51
33	Harry	Truman	Missouri	1884	61
34	Dwight	Eisenhower	Texas	1890	63
35	John	Kennedy	Massachusetts	1917	44
36	Lyndon	Johnson	Texas	1908	55
37	Richard	Nixon	California	1913	56
38	Gerald	Ford	Nebraska	1913	61
39	James	Carter	Georgia	1924	53
40	Ronald	Reagan	Illinois	1911	70
			Average Age At Inauguration		55.55
			Std Dev At Inauguration		6.066094

To keep our following examples simple, the following examples will use rounded averages of 55 and a rounded standard deviation of 6.

Distribution

Not only do we need to know the center point and the magnitude of the spread, we also need to know where the spread is occurring.

The distribution answers the question, "Where Is The Spread?". From what we saw of the spread of the people's weights, the best way to analyze the distribution is through pictures. This is accomplished through either a frequency table or histogram. The **frequency table** is the tabular means for analyzing where the data is distributed. The **histogram** is the pictorial technique for analyzing the distribution.

The frequency table counts the items in the data that fall into a series of intervals. The example shown below includes the age at election of the first 40 presidents of the United States. The histogram then shows them in a pictorial manner.

The Age Groups	The Number of Presidents
40	0
45	2
50	6
55	12
60	11
65	6
70	4
75	0

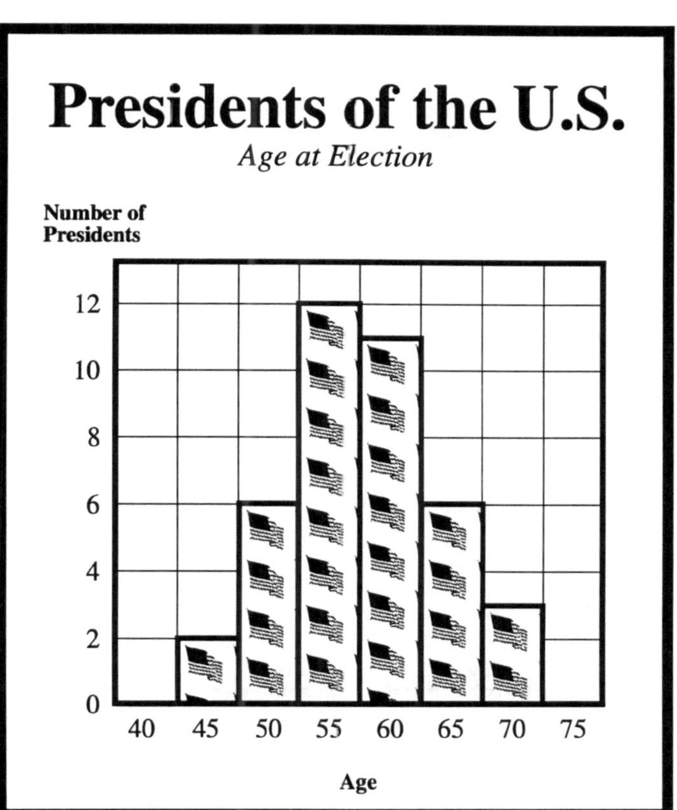

The frequency table is constructed by determining the following items:

X-Axis Range
Number of Bins
Width of each Bin
Number of data points in each bin

Frequency Table Construction

The steps below show us how to prepare a frequency table. The frequency table uses cells to count the number of occurrences for each group. These groups may be referred to as classes, groups, or bins.

STEPS

1. Determine the range of the X-Axis
   a. Determine and record the lowest actual value. For our example, the youngest president is Theodore Roosevelt whose age was 43 at inauguration.
   b. Determine and record the highest actual value. For our example, the oldest president is Ronald Reagan whose age was 70 at inauguration.
   c. Determine and record the lowest contract or specification value. For our example, the Constitution states no one shall be elected to the office of president younger than 35 years of age.
   d. Determine and record the highest contract or specification value. For our example, there is no upper end to the age of a president.
   e. Now, compute the low end of the X-Axis by taking the lowest actual value or specification. Should no low specification be available, then provide some space on the low axis for any future values outside our current values. The lowest of the actual (43 years of age) for our presidents and the low specification from the Constitution (35 years of age) is now found. This value of 35 years of age becomes the low end of the X-axis.
   f. Now, compute the high end of the X-Axis by taking the highest actual value or specification. Should no upper specification be available, then provide some space on the high axis for any future values outside our current values. The highest of the actual (70 years of age) for our presidents and the low specification is not specified in the Constitution. The extra space for future high values is 75 years of age and now becomes the high end of the X-axis.
   g. Finally, the range of the X-Axis is computed and saved for the next step.

	Actual	Specification	X-Axis Boundary
Minimum	43	35	35
Maximum	70	None	75
		Range	**40**

2. Number of classes.
   a. Determine the number of classes. A good rule of thumb is approximately seven to ten classes. Too few or too many bins will cause the distribution to have an inappropriate appearance. For the example of the presidents, we will select eight classes because the range of 40 is evenly divisible by eight.

3. Class Width
   a. To determine the bar width for the histogram of the classes, divide the X-Axis range by the number of classes as selected from above. For the presidents, the range of 40 years is divided by the number of classes which is eight. Each class will then be five years wide.

4. Prepare the frequency table:
   a. In a column, enter the series of values defining the lower limits of the classes. The first class will start at 35 years of age for a length of five years. The second class will start at 40 for a range of five years. Continue until all the classes are identified as shown below.
   b. Leave a blank column to the right of the bin values, including one blank cell below the largest value. This column will contain the frequency distribution. The table below shows how the list bins would look for the presidents of the United States.

The Age Groups	The Number of Presidents
35-40	
40^{+}-45	
45^{+}-50	
50^{+}-55	
55^{+}-60	
60^{+}-65	
65^{+}-70	
70^{+}-75	

5. Count the number of occurrences for each bin. As an example, George Washington's age of 57 when inaugurated into office falls in the 55-60 class, John Adams' age of 62 falls in the 60-65 range, and Thomas Jefferson's age of 58 would add a second value with Washington in the 55-60 class. The counting would continue through our list to Ronald Reagan.

The Age Groups	The Number of Presidents
35^{+}-40	0
40^{+}-45	2
45^{+}-50	6
50^{+}-55	12
55^{+}-60	11
60^{+}-65	6
65^{+}-70	4
70^{+}-75	0

How can spreadsheets build frequency tables and histograms?

The steps below show us how to prepare a frequency table or a Histogram. Should you have a spreadsheet like *Lotus* or *Excel* we could use these steps to prepare the bar chart called Histogram.

STEPS for Lotus

1. Prepare the distribution table (Bin Range):
   a. In a column, enter a series of values defining the lower limits of the bins for the desired distribution.
   b. Leave a blank column to the right of the bin values, including one blank cell below the largest value. This column will contain the frequency distribution.

2. */ Data Distribution*

3. Specify the Values range:
   a. Move the cell pointer to one corner of the Values Range. Press . to anchor the cell pointer, highlight the range, and press RETURN.

4. Specify the Bin Range:
   a. Move the cell pointer to one corner of the Bin Range. Press . to anchor the cell pointer, highlight the range, and press RETURN.

STEPS for Excel

1. Prepare the distribution table (Bin Range):
   a. In a column, enter a series of values defining the lower limits of the bins for the desired distribution.
   b. Leave a blank column to the right of the bin values, including one blank cell below the largest value. This column will contain the frequency distribution.

2. Select
   a. *Tools*
   b. *Analysis* (this is an Excel add-in and may need to be installed)
   c. and then *Histogram*

3. Specify the Values range:
   a. Move the cell pointer to one corner of the Values Range. Press . to anchor the cell pointer, highlight the range, and press RETURN.

4. Specify the Bin Range:
   a. Move the cell pointer to one corner of the Bin Range. Press . to anchor the cell pointer, highlight the range, and press RETURN.
5. Select *Charts*

Histogram

The Histogram displays in pictorial form the measurement distribution and explains where the spread or *concentration of values* is occurring. The horizontal axis shows how the measurement groups were formed. The vertical axis depicts the number of occurrences for each group.

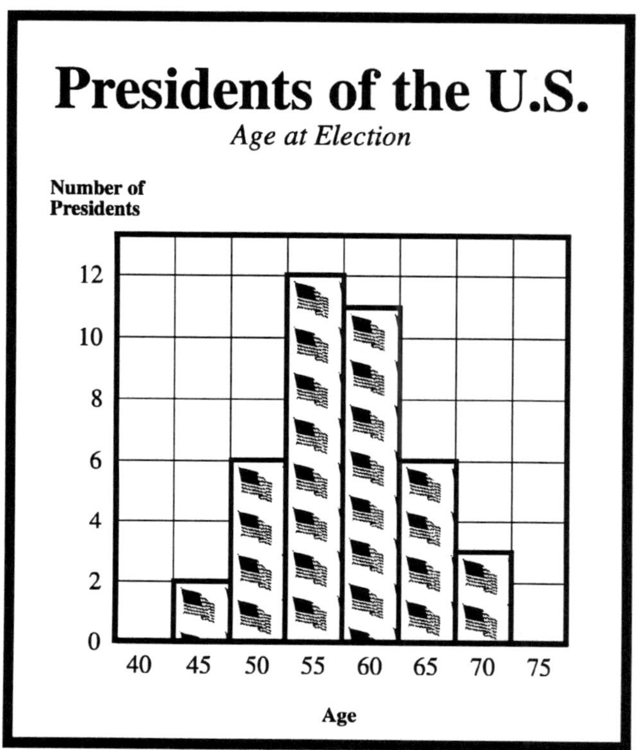

The vertical or ***Y-Axis*** shows the number of events that occurred for this group. The age group from 35 years to 40 years has no presidents, 40 years to 45 years has two presidents, while 45 years to 50 years has six presidents. This chart continues for all the groups.

The horizontal or **X-Axis** shows how the groups are formed in five year increments.

This chart is used to assess how well the distribution is fitted to a normal curve. The location of the spread of the United States presidents' ages is shown in the histogram above. From our prior statistics, we know the average age of the presidents is 55 years of age. The bars that form the histogram are the tallest closest to this average. As the ages become older and younger further from the average, the number of occurrences drops rapidly on both sides of the average. The bars above the average and below the average are symmetric to each other. From this analysis, we can conclude that presidents' ages are normally distributed and thus we can use the probabilities of the normal to make predictions. Now we need to learn the normal probabilities so that we can use them for predictions.

Normal Distribution Everyone has been graded on a bell shaped curve at some point in school. This bell curve is another name for our friend, the **normal distribution**. The normal distribution is used to convert the actual grades or test scores to a letter grade.

Description Of The Normal Distribution

Average ...is the central point of the measurements.
Standard Deviation ...is the variability of the measurements.

The distribution is symmetric about its average. Fifty percent of the data points are above the average and fifty percent of the measurements are below the average. The frequency of occurrence (or the number of occurrences) drops rapidly as we move further and further away from the center. This characteristic is true both above and below the average. The frequency of occurrences above and below the average are symmetric.

The chart to the right is used to depict a picture of how a process with its measurements will look. The chart shows the spread of grades from an English Literature class of thirty students. When the score measurements are converted to grades and look like the diagram, the process is well represented by a normal distribution. The perfect normal does not have the steps of a histogram but is a smooth curve.

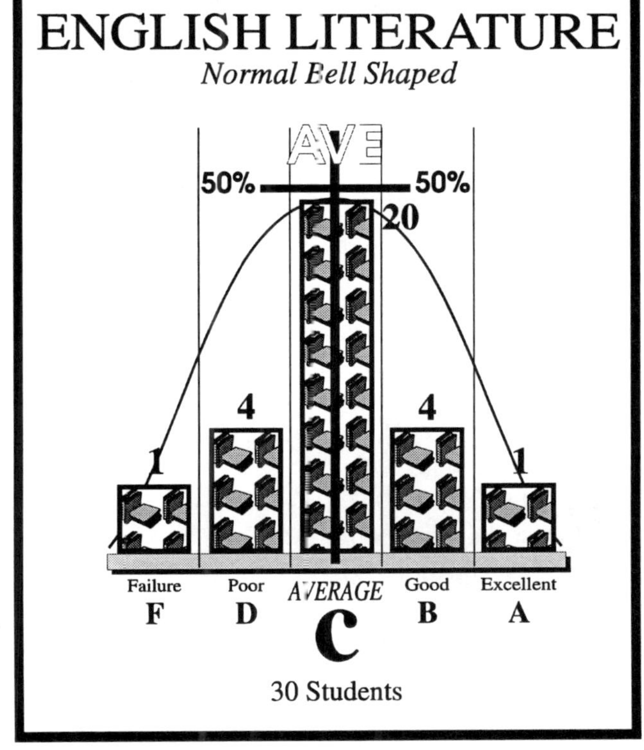

Why is the Normal Distribution important? Having data that is well represented by a normal distribution allows us to predict using the probabilities of a normal. Because a good understanding of these probabilities is essential, let's now look at these probabilities in more detail.

This chart show how a typical class in English literature could have been graded using a bell shaped curve. Out of a class of thirty students twenty would receive a "C", four would receive a "B", another four would receive a "D", one would receive an "A", and one would receive an "F".

How Do We Determine the Normal Probability?

All measurements that track like a normal distribution have a very definitive set of probabilities. Since the normal distribution is symmetric, the probabilities above and below the average are exactly the same.

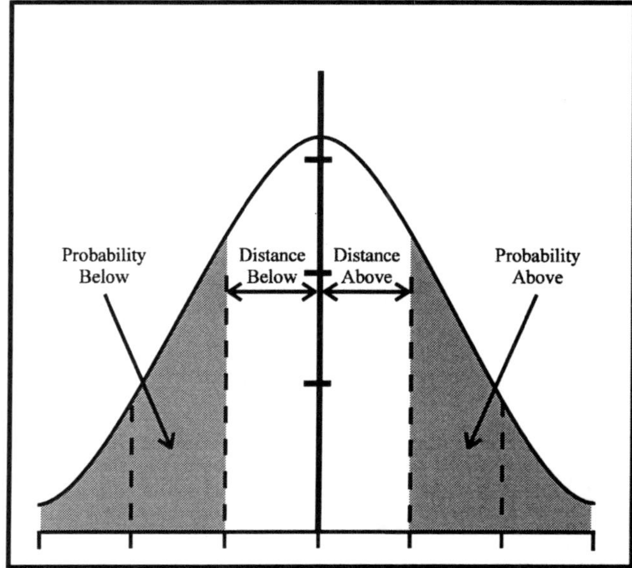

As the distance away from the mean, or center, of normal distribution gets bigger, the probability gets smaller. This distance can be calculated in the following formula. Due to the normal distribution's symmetry, only one formula is required and the sign is always positive.

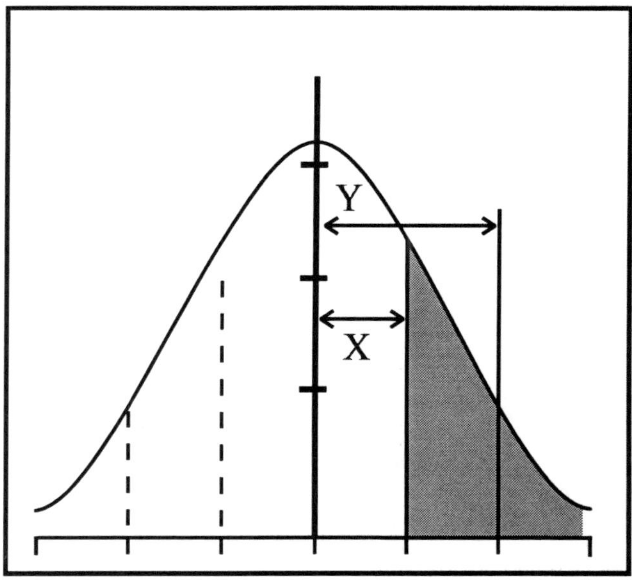

Distance from the average $= X_i - X_{Ave}$

Should we desire to compute the probability of a person's presidential election based on age, this formula is used. For a person 61 years of age or older, we can compute this distance by subtracting the average from the age.

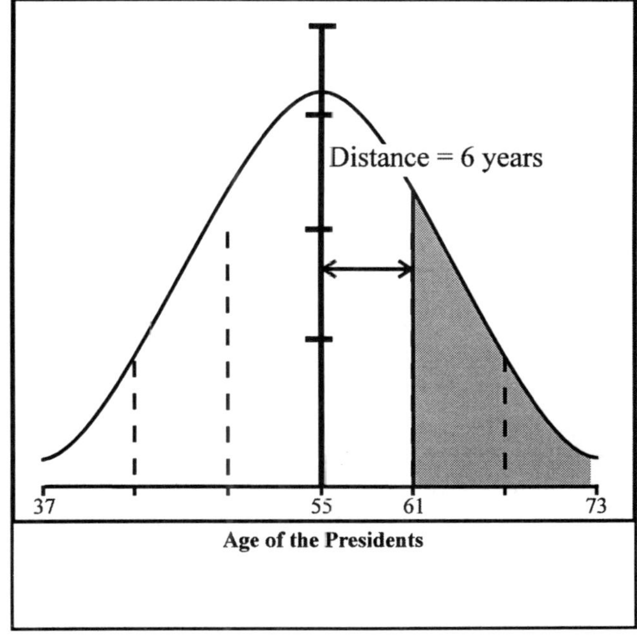

61 years Distance from the average $= X_i - X_{Ave}$

$= 61-55$

$= 6$ years of age

Now to make for a general rule for *all measurements* we must convert this distance to a **ratio**. The ratio will be the distance as a function of variability. This ratio will remove units of measure such as the president's age and will give us a conversion factor that all measurements can use. The ratio's formula is calculated below. The ratio of the distance from the average divided by the standard deviation is called **Z**.

$$Z = \left| \frac{X_{Ave} - X_i}{STD} \right|$$

To compute the Z value for our example of 61 years and older, the following must be computed:

$$Z_{61} = \frac{55-61}{6}$$

$$= \frac{6}{6}$$

$$= 1$$

Once again Z is the ratio of the distance away from the average as a function of variability measured by standard deviation. This Z value, which is a ratio, allows the use of one probability table for any measurement. Now this table is used to locate the probability for the particular Z value that has been computed. For our example, the Z-value is one.

Z	X.X0
4	.00003
3	.00135
2	.0228
1	**.1587**
0	.500

For a person 64 years of age or older, we can compute this distance by subtracting the average from the age.

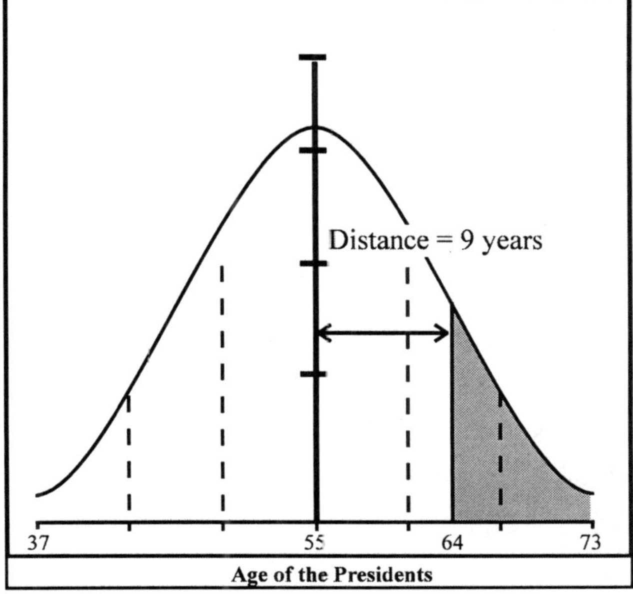

64 years Distance from the average $= X_i - X_{Ave}$

$= 64-55$

$= 9$ years of age

This ratio once again will remove all the units of measure such as the president's age. This ratio will give us a conversion factor for 64 years that we can use to arrive at the probability. The ratio for 64 years is calculated below.

$$Z = \left| \frac{X_{Ave} - X_i}{STD} \right|$$

To compute the Z value for this example of 64 years and older, the following must be computed:

$$Z_{64} = \frac{55-64}{6}$$

$$= \frac{9}{6}$$

$$= 1.5$$

Once again Z is the ratio of the distance away from the average as a function of variability measured by standard deviation. This Z value, which is a ratio, allows the use of one probability table for any measurement. Now the table is expanded to include more detail with Z-values to one decimal place. We look down the Z-value column until we find the value of 1.5. The probability for the particular Z value that has been computed, and for our example, is immediately to the right. The probability for Z of 1.5 is .0668. In other words, 6.68 percent of the presidents would be 64 and older.

Z	X.X0
4.0	.00003
3.5	.00023
3.0	.00135
2.5	.0062
2.0	.0228
1.9	.0287
1.8	.0359
1.7	.0446
1.6	.0548
1.5	**.0668**
1.4	.0808
1.3	.0968
1.2	.1151
1.1	.1357
1.0	.1587
.9	.1841
.8	.2119

For a person 65 years of age or older, we can compute this distance by subtracting the average from the age.

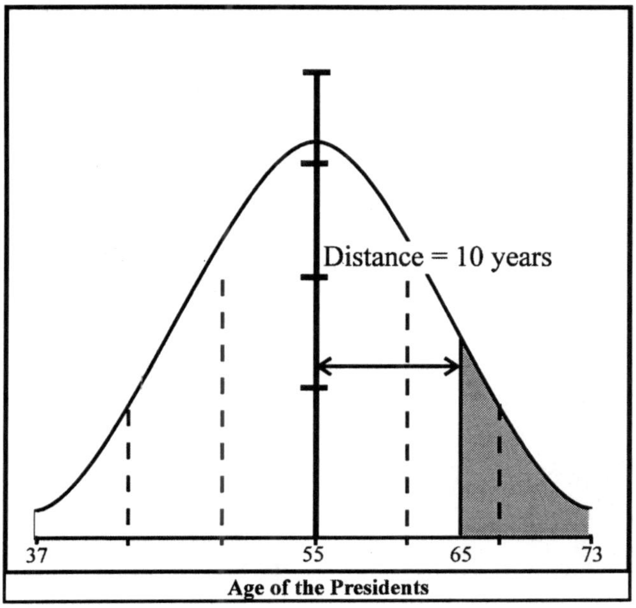

Distance = 10 years
37 55 65 73
Age of the Presidents

65 years Distance from the average	=	$X_i - X_{Ave}$
	=	65-55
	=	10 years of age

To remove the age units of measure, the Z-ratio will once again be used. The Z-ratio will give us a conversion factor for 65 years that we can use to arrive at the probability. The ratio for 65 years is calculated below.

$$Z = \left| \frac{X_{Ave} - X_i}{STD} \right|$$

To compute the Z value for this example of 65 years and older, the following must be computed:

$$Z_{65} = \frac{55-65}{6}$$

$$= \frac{10}{6}$$

$$= 1.67$$

To take the Z value to a second decimal place, the normal table must be revised. Now a series of columns are used to identify the second decimal place. The following shows the look of the table. The first column identifies the Z-value whole and first decimal place and the remaining columns identify the second decimal place. For a Z-Value of 2.93, the row would be 2.9 and the column would X.X3 with the resulting probability being .0017.

Z	X.X0	X.X1	X.X2	X.X3	X.X4	X.X5	X.X6	X.X7	X.X8	X.X9
3.0	.00135	.00131	.00126	.00122	.00118	.00114	.00111	.00107	.00104	.00100
2.9	**.0019**	**.0018**	**.0018**	**.0017**	.0016	.0016	.0015	.0015	.0014	.0014
2.8	.0026	.0025	.0024	.0023	.0023	.0022	.0021	.0021	.0020	.0019

Now turn to the next page and look up the Z_{65} of 1.67. Find the row 1.6 and the column X.X7 which should give a probability value of .0475. Thus we would expect 4.75 percent of the presidents to be 65 and older. The reverse of 65 and younger yields 95.25 percent of the presidents falling in this range. The next page provides a full normal distribution table for Z-values to two decimal places.

Remember, the first column and top row are the Z-values. The remaining body of the table is composed of probabilities for the corresponding Z-values.

Normal Distribution Table The table below allows you to look up the area (percent) of the normal distribution probability based on the distance away from the center of the distribution monitored by the average. This distance away from the average as a function of the variability monitored by standard deviation is a ratio called the **Z-value**. This Z-value can then be used to look up the probability of occurrence outside the point. (*Probability* refers to the shaded area in the graph.) The table below is a **tail-oriented probability table** with the values representing one of the shaded areas in the diagram. This probability is the same for both upper and lower tails because of the symmetry of the normal distribution. The first column represents the Z-value to the whole number and one decimal place while the top row represents the second decimal place for the Z-value.

$$Z = \left| \frac{X_{Ave} - X_i}{STD} \right|$$

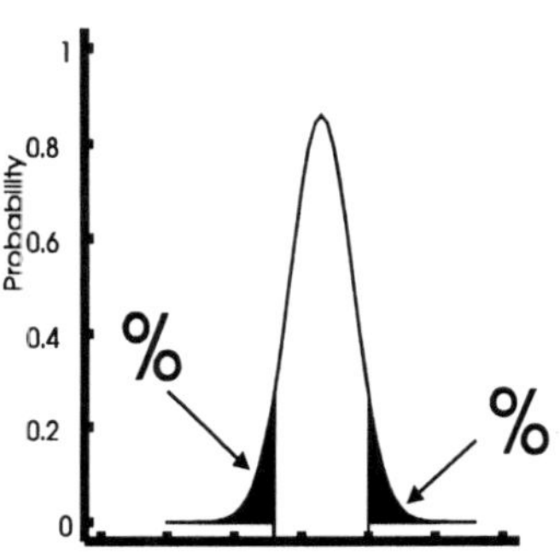

Z	X.X0	X.X1	X.X2	X.X3	X.X4	X.X5	X.X6	X.X7	X.X8	X.X9
4.5	0.000003	0.000003	0.000003	0.000003	0.000003	0.000003	0.000003	0.000002	0.000002	0.000002
4.0	0.000032	0.000030	0.000029	0.000028	0.000027	0.000026	0.000025	0.000024	0.000023	0.000022
3.5	0.000233	0.000224	0.000216	0.000208	0.000200	0.000193	0.000185	0.000179	0.000172	0.000165
3.0	0.001350	0.001306	0.001264	0.001223	0.001183	0.001144	0.001107	0.001070	0.001035	0.001001
2.9	0.001866	0.001807	0.001750	0.001695	0.001641	0.001589	0.001538	0.001489	0.001441	0.001395
2.8	0.002555	0.002477	0.002401	0.002327	0.002256	0.002186	0.002118	0.002052	0.001988	0.001926
2.7	0.003467	0.003364	0.003264	0.003167	0.003072	0.002980	0.002890	0.002803	0.002718	0.002635
2.6	0.004661	0.004527	0.004397	0.004269	0.004145	0.004025	0.003907	0.003793	0.003681	0.003573
2.5	0.006210	0.006037	0.005868	0.005703	0.005543	0.005386	0.005234	0.005085	0.004940	0.004799
2.4	0.008198	0.007976	0.007760	0.007549	0.007344	0.007143	0.006947	0.006756	0.006569	0.006387
2.3	0.010724	0.010444	0.010170	0.009903	0.009642	0.009387	0.009137	0.008894	0.008656	0.008424
2.2	0.013903	0.013553	0.013209	0.012874	0.012545	0.012224	0.011911	0.011604	0.011304	0.011011
2.1	0.017864	0.017429	0.017003	0.016586	0.016177	0.015778	0.015386	0.015003	0.014629	0.014262
2.0	0.022750	0.022216	0.021692	0.021178	0.020675	0.020182	0.019699	0.019226	0.018763	0.018309
1.9	0.028716	0.028067	0.027429	0.026803	0.026190	0.025588	0.024998	0.024419	0.023852	0.023295
1.8	0.035930	0.035148	0.034379	0.033625	0.032884	0.032157	0.031443	0.030742	0.030054	0.029379
1.7	0.044565	0.043633	0.042716	0.041815	0.040929	0.040059	0.039204	0.038364	0.037538	0.036727
1.6	0.054799	0.053699	0.052616	0.051551	0.050503	0.049471	0.048457	0.047460	0.046479	0.045514
1.5	0.066807	0.065522	0.064256	0.063008	0.061780	0.060571	0.059380	0.058208	0.057053	0.055917
1.4	0.080757	0.079270	0.077804	0.076359	0.074934	0.073529	0.072145	0.070781	0.069437	0.068112
1.3	0.096801	0.095098	0.093418	0.091759	0.090123	0.088508	0.086915	0.085344	0.083793	0.082264
1.2	0.115070	0.113140	0.111233	0.109349	0.107488	0.105650	0.103835	0.102042	0.100273	0.098525
1.1	0.135666	0.133500	0.131357	0.129238	0.127143	0.125072	0.123024	0.121001	0.119000	0.117023
1.0	0.158655	0.156248	0.153864	0.151505	0.149170	0.146859	0.144572	0.142310	0.140071	0.137857
0.9	0.184060	0.181411	0.178786	0.176186	0.173609	0.171056	0.168528	0.166023	0.163543	0.161087
0.8	0.211855	0.208970	0.206108	0.203269	0.200454	0.197662	0.194894	0.192150	0.189430	0.186733
0.7	0.241964	0.238852	0.235762	0.232695	0.229650	0.226627	0.223627	0.220650	0.217695	0.214764
0.6	0.274253	0.270931	0.267629	0.264347	0.261086	0.257846	0.254627	0.251429	0.248252	0.245097
0.5	0.308538	0.305026	0.301532	0.298056	0.294598	0.291160	0.287740	0.284339	0.280957	0.277595
0.4	0.344578	0.340903	0.337243	0.333598	0.329969	0.326355	0.322758	0.319178	0.315614	0.312067
0.3	0.382089	0.378281	0.374484	0.370700	0.366928	0.363169	0.359424	0.355691	0.351973	0.348268
0.2	0.420740	0.416834	0.412936	0.409046	0.405165	0.401294	0.397432	0.393580	0.389739	0.385908
0.1	0.460172	0.456205	0.452242	0.448283	0.444330	0.440382	0.436441	0.432505	0.428576	0.424655
0.0	0.500000	0.496011	0.492022	0.488033	0.484047	0.480061	0.476078	0.472097	0.468119	0.464144

Normal Curve Probabilities Review

The area under the curve represents the probability of occurrence.

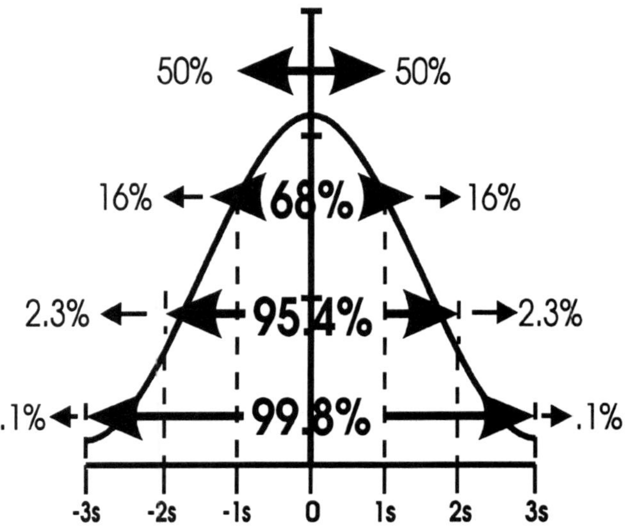

What does a Standard Deviation mean? Write the probability of occurrence at various spots on the X-axis in the blanks below.

Standard Deviation	Left	Right	Both Sides
1	_____	_____	_______
2	_____	_____	_______
3	_____	_____	_______

An occurrence outside plus or minus one standard deviation (one is the z-Value) equates to a probability of 16 percent above or 16 percent below or 68 percent inside. An occurrence outside plus or minus two standard deviation (two is the z-Value) equates to a probability of 2.3 percent above or 2.3 percent below or 95.4 percent inside. An occurrence outside plus or minus three standard deviation (three is the z-Value) equates to a probability of .1 percent above or .1 percent below or 99.8 percent inside.

Returning to the English literature class, how were the different letter grades determined? First the "C" grades were inside the range of plus or minus one standard deviation. In this range, the probability is 68 percent. Thus 68% of 30 students is the twenty students receiving the C grade.

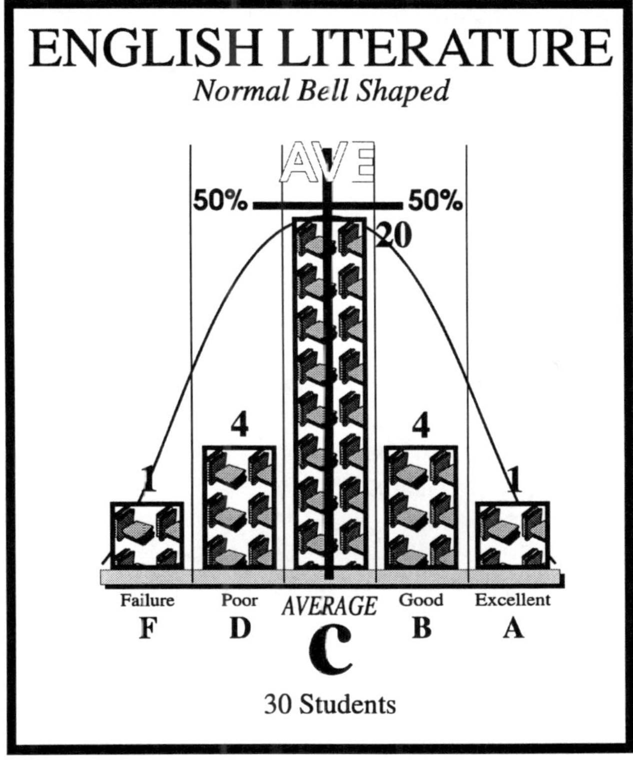

Notice that 95.4% of the grades fall in the plus or minus two standard deviations. Of these, 68% have already received their C grade leaving a total of 27% receiving a B or a D. Since the normal distribution is symmetric, half the students and those above the average will receive a B which is 13% or four B grades. The remaining four students in the low end of this range will receive a D grade.

Now 99.8% of the grades fall in plus or minus three standard deviations. Similarly, 95.4% have already received either an A, B, or C. The remaining 4.4%, or two students, will receive an F or A grade. The student on the low side of the average will receive the F while the student on the high side of the average will receive the A.

Using the Normal Table for Information Risk Management

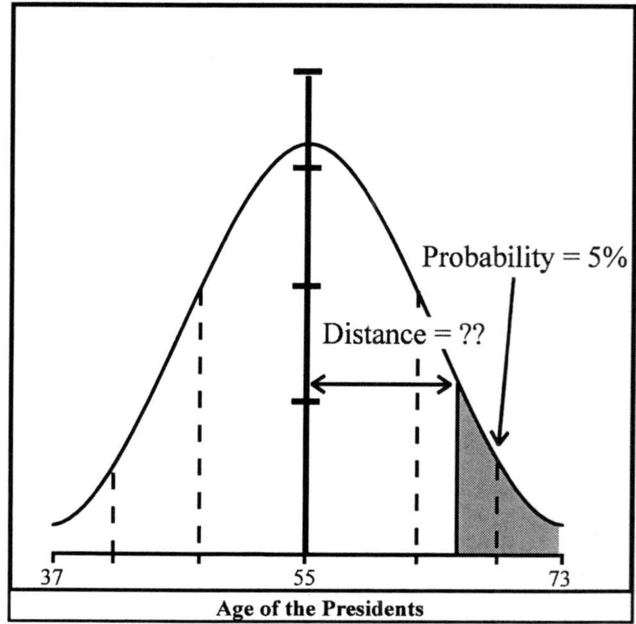

Not only can the normal table be used to compute probability, but this process can be reversed to have a probability of risk and arrive at the spots where this amount of risk would occur. The English literature class is an example of having known probabilities and arriving at the number of occurrences or the particular spot. The Z- value formula below remains.

$$Z = \left| \frac{X_{Ave} - X_i}{STD} \right|$$

Now the equation is simply rearranged using algebra to solve for X_i. The resulting equation becomes:

$$X_i = X_{Ave} \pm z * Std$$

For example, compute the age of presidents where the risk above average is five percent and the risk below the average is five percent. First the Z associated with 5% risk must be determined. Five percent equates to .05. Now we go to the normal table and look in the body for the value closest to .05. On the row 1.6, the values of .0505 and .0495 are found. The probability of .0505 has a Z of 1.64 while the probability of .0495 has a Z of 1.65. Since .05 is in the center of the two probabilities, the Z in the center of 1.64 and 1.65 or 1.645 is selected.

Now the average age of the presidents of 55 years and the standard deviation of the presidents of 6 years are both used in conjunction with the selected Z of 1.645 to compute the selected spots.

$$\begin{aligned} X_i &= X_{Ave} \pm z * Std \\ &= 55 \pm 1.645 * 6 \\ &= 55 \pm 9.87 \\ &= 45.13 - 64.87 \end{aligned}$$

Thus five percent of the presidents should be over 64.87 years of age and five percent of the presidents should be under 45.13 years of age.

Basic Statistics Summary

For proper numerical analysis technique, we must monitor each of the following:

- Central Tendency
- Variability
- Distribution

Failure to monitor all three will lead to catastrophic mistakes.

MONITOR	**FORMULA**

Central Tendency or balance point of the data

Average $$\frac{\sum_{i=1}^{N} Xi}{N}$$

Variability or amount or magnitude of the data's spread

Range Maximum-Minimum

Standard Deviation $$\sqrt{\frac{\sum_{i=1}(X_{AVE} - X_i)^2}{N-1}}$$

Distribution or location of the data's spread

Frequency Table

Histogram

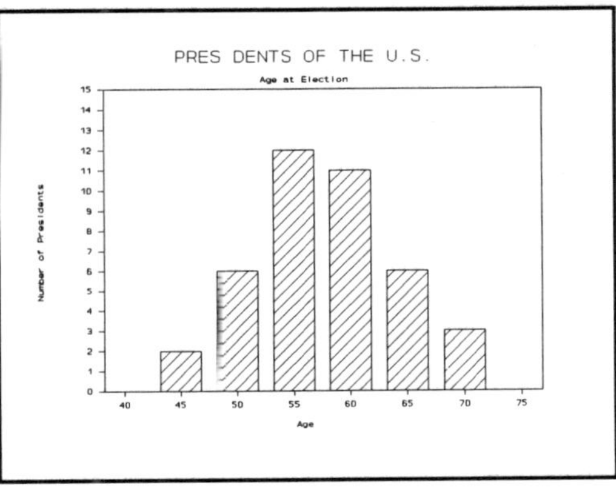

If the data that we are analyzing is normally distributed, then the probabilities associated with the normal curve can be used to make predictions. These predictions use a normal table. This normal table is comprised of two values: the Z-value and the probability.

If the probability is to be determined, then the Z-value formula below must be used. Our knowns must be the average of the data, the standard deviation of the data, and the point that we desire to know the probability.

$$Z = \left| \frac{X_{Ave} - X_i}{STD} \right|$$

If the probability is known, then the equation is simply rearranged using algebra to solve for X_i. The resulting equation is shown below. The knowns of the average of the data, the standard deviation of the data, and the risk that we are prepared to take must be determined.

$$X_i = X_{Ave} \pm z * Std$$

We have discussed the first major flaw associated with a statistical analysis: improper technique. An understanding of the importance of proper technique and how to do a proper study should now be in place. Before we go to the other major flaw of improper grouping, a clear understanding of some math tricks that make SPC work will be covered in the next chapter.

Problem

From our analysis of the electricity usage, compute the amount of funds that must be on reserve to pay the maximum kilowatt usage for any one month. This study will assume that the distribution of the kilowatts is normal, the cost of a kilowatt is $.10, and that we are prepared to take a 5% risk of not having enough funds available to cover the usage. Use the space below to show all your work and the amount of money that you must keep in reserve. The solution to this problem and all later problems is found prior to the appendices.

Chapter Two: Math Tricks to Make SPC Work

Statistical Process Control Flow

Statistical Process Control (SPC) is used to analyze and monitor a continuous process. SPC should be used to monitor both product and process characteristics of all continuous operations.

SPC is comprised of a tactical tool called a **control chart** and a strategic tool called the **capability study**. Control charts are the operator's tool to run the machine while the capability study is management's tool to assess the impact on the customer.

Later in this book we will discuss in depth the SPC flow to our left. For now we must focus on the math that supports SPC.

Control Charts: A control chart is the alarm system to note changes in a continuous operation. A control chart is a pictorial alarm system to detect any change in the consistency of the machine's operation. Any deviations from the consistent operation are a change and must be investigated. Remember, change can be either an improvement or a deterioration. Another way of describing a control chart is to call it a *change alarm system.* The objective of a control chart is to provide a tool that helps the operator run the process to its optimum.

Capability Study: A capability study is the tool to assess the impact of all changes detected by a control chart. The capability study is the management tool to analyze the change and determine whether it is an improvement or a deterioration.

Since control charts only detect changes and capability studies only assess the impact, then specialized mathematics can be used for each purpose. In the studies of process management, basic principles must be clearly understood. The second principle allows all the mathematics to work in SPC. The second principle is stated below.

Second Principle of Process Management (Division of Labor) drives the decision making process. It dictates who will be in charge of the different kinds of decisions (strategic or tactical). We are doomed to a state of quality hell if the managers of the future fail to fully grasp Division of Labor.

Second Principle of Process Management

Division of Labor is the framework for all aspects of decision making. It must be clearly understood to separate the strategic and tactical decisions. Operations makes the tactical decisions of running the facility. Management makes the strategic decisions of assessing the facility's suitability for the job.

Control is a major tactical issue associated with operational issues. Capability is a major strategic issue associated with management, customer, and sales issues.

If we have a clear understanding of the Second Principle of Process Management, or this "Division of Labor", we will possess the key to appreciating the difference between tactical and strategic decisions.

Division of Labor requires a clear understanding between management/customer/sales issues and operational issues.

Tactical The art or skill of employing *available* means to accomplish an end.

Operational Decisions are tactical and detect change. Operations or tactics use the available means. All tactical analysis must support this mission.

Strategic The art of devising or employing plans or stratagems toward a goal.

Management Decisions are strategic and assess impact. Management or strategy selects, assesses, and provides the means. All strategic analysis must support this mission.

Both decision types are crucially important. The two sets of decisions have a very specific set of mathematics principles that work only for either tactical or strategic **but never** for both. Very specialized tools must be provided to support each decision type specifically.

These mathematical tools will provide a common language for discussing process deviations, performance, and change. The tools must be used following the classic definition to avoid confusion.

CAUTION ! Do not confuse control charts with strategic management reports that use customer specifications.

Division of Labor allows us to have two sets of books for very precise objectives and to serve very different audiences. In the accounting community, the same set of numbers may be manipulated several different ways to accomplish differing needs. The financial information required to support a cash flow projection is completely different from the needs of a profit and loss statement sent to the government for tax purposes. Our tactical and strategic objectives require different information, thus our need to discuss the two statistical sets of books.

Let's use our newly learned basic statistical skills to analyze the set of numbers shown below. I have derived the numbers to create a problem for our new skills. From Chapter One, you learned about the normal distribution. The normal distribution allows us the means of using a probability table to make projections. These projections can only be used if the distribution is normal.

The chart below shows a set of evenly distributed numbers that range from zero to one hundred. Each number in the range has a equal chance of occurring for each value. The chart shows two sets of numbers and each set contains five columns.

==============Random Numbers=================									
31	94	7	0	1	87	95	63	84	93
78	86	5	51	15	90	5	5	75	96
74	76	65	88	52	95	44	75	69	77
62	49	87	46	89	68	44	54	37	15
11	3	92	91	79	71	12	92	1	36
76	23	36	1	83	36	79	51	38	42
31	64	93	30	87	13	90	58	97	93
93	39	65	20	63	33	59	18	45	49
40	54	85	11	71	18	63	50	99	70
78	81	11	7	27	75	8	72	85	80
90	21	10	21	43	3	68	12	26	97
6	90	78	52	48	93	83	20	62	80
97	94	50	56	91	18	17	15	45	21
26	23	25	95	83	37	41	34	26	41
52	27	52	64	65	91	34	44	82	99
17	8	5	86	93	40	27	5	78	14
18	12	39	49	36	27	77	49	71	47
8	23	94	32	13	56	38	71	90	42
78	44	60	27	10	53	24	55	60	98
0	59	100	38	66	10	48	77	5	89
60	97	85	82	42	100	42	95	51	98
3	30	64	56	18	8	83	19	92	78
79	85	95	29	38	1	28	26	77	95
30	98	84	92	15	13	98	44	27	3
15	83	12	57	63	57	99	92	99	83
64	59	25	40	58	24	10	73	24	70
47	12	7	35	89	78	27	69	72	80
12	26	68	38	99	93	82	34	88	67
30	80	91	16	60	41	28	10	2	42
55	26	48	85	36	1	57	14	69	96
81	30	8	31	12	47	58	40	29	61
98	9	99	64	11	98	71	74	20	71

From Chapter One we must compute a monitor of central tendency, variability, and the data distribution. The monitor of central tendency that I want you to compute is the average. The monitor of variability is standard deviation. The monitor of the data distribution is a histogram.

	Individual	
Central Tendency	Average	52
Variability	Standard Deviation	30
Distribution	Histogram	Non-Normal

The average of the data is 52. The standard deviation is 30. The distribution of the data is a very flat distribution and is not even near a normal distribution. Thus we must conclude that this set of values is non-normal.

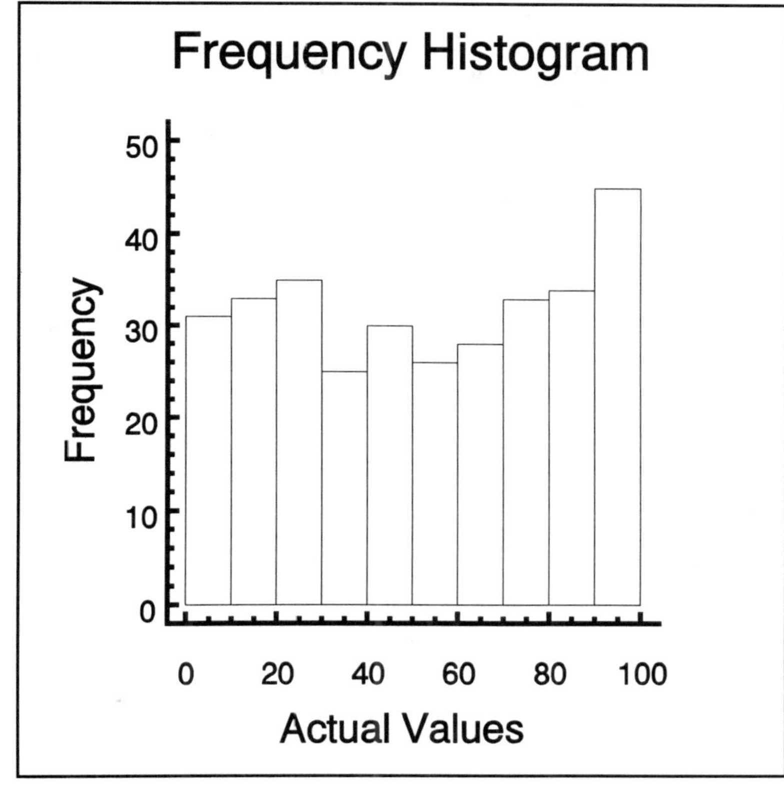

In the real world, many processes, products, or data sets are not always well represented by a normal distribution.

The above must be used by management for strategic calculations. Since the distribution is not bell in shape, a much more sophisticated analysis must be used. This analysis must include the computing of the probabilities for this one particular set of numbers. Remember this analysis will compare the product that we are making to the product the customer desires. In this analysis, customer specifications will be included.

Tactical Support To support operations, clarity of Division of Labor is required. The operator's role is to correctly and consistently run the machine. The operator must run the machine to its optimum. Customer specifications cannot be used to run a machine let alone get the machine to its optimum. With this objective clear, then we can introduce a natural phenomenon known as **central limit theorem**.

Normalized Data

There is a natural set of factors which tend to make these processes have a normal distribution. This is explained by the **central limit theorem**. This theorem implies that the sum of a large number of data points will be normally distributed regardless of the distribution of the individual data points, and that the mean of n samples will be normally distributed.

This process is a means of normalizing a non-normal process. The process is accomplished by averaging subgroups of data points.

We compute the average of our first set of five values from the table: 31, 94, 7, 0, and 1, as shown below:

$$\begin{aligned} Average &= \frac{31+94+7+0+1}{5} \\ &= 26.6 \end{aligned}$$

Similarly, we compute the average of our second set of five values from the table: 78, 86, 5, 51, and 15, as shown below:

$$\begin{aligned} Average &= \frac{78+86+5+51+15}{5} \\ &= 47.0 \end{aligned}$$

Again, we compute the average of our third set of five values from the table: 74, 76, 65, 88, and 52, as shown below:

$$\begin{aligned} Average &= \frac{74+76+65+88+52}{5} \\ &= 71.0 \end{aligned}$$

Once more, we compute the average of our fourth set of five values from the table: 62, 49, 87, 46, and 89, as shown below:

$$\begin{aligned} Average &= \frac{62+49+87+46+89}{5} \\ &= 66.6 \end{aligned}$$

Still again, we compute the average of our fifth set of five values from the table: 11, 3, 92, 91, and 79, as shown below:

$$Average = \frac{11+3+92+91+79}{5}$$
$$= 55.2$$

Finally, we compute the average of our sixth set of five values from the table: 76, 23, 36, 1, and 83, as shown below:

$$Average = \frac{76+23+36+1+83}{5}$$
$$= 43.8$$

These computations are continued for every set of values. Each set of five values is called a **subgroup**. The table below shows the results of the normalizing of the first six subgroups with their normalized averages.

======Random Numbers==					Row Average
31	94	7	0	1	26.6
78	86	5	51	15	47.0
74	76	65	88	52	71.0
62	49	87	46	89	66.6
11	3	92	91	79	55.2
76	23	36	1	83	43.8

This process is continued for all the subgroups in our set of values. The values below are the resultant normalized averages for every one of the subgroups of five. The first value of 26.6 in the table is the average of five individual values in the first group. Each value in the table is its corresponding subgroup's average.

Average of the Random Numbers	
26.6	84.4
47.0	54.2
71.0	72.0
66.6	43.6
55.2	42.4
43.8	49.2
61.0	70.2
56.0	40.8
52.2	60.0
40.8	64.0
37.0	41.2
54.8	67.6
77.6	23.2
50.4	35.8
52.0	70.0
41.8	32.8
30.8	54.2
34.0	59.4
43.8	58.0
52.6	45.8
73.2	77.2
34.2	56.0
65.2	45.4
63.8	37.0
46.0	86.0
49.2	40.2
38.0	65.2
48.6	72.8
55.4	24.6
50.0	47.4
32.4	47.0
56.2	66.8

Now what can these numbers do for us? First let's analyze the subgroup averages using the proper techniques that we learned in Chapter One.

Normalized Data This process is a means of normalizing a non-normal process. The process is accomplished by averaging subgroups of data points. Then the averages for each subgroup must be analyzed. This set of tools is specifically a tactical tool to detect change. The subgroup averages from our example are now ready for analysis. Note the results of the simple statistics we learned in Chapter One.

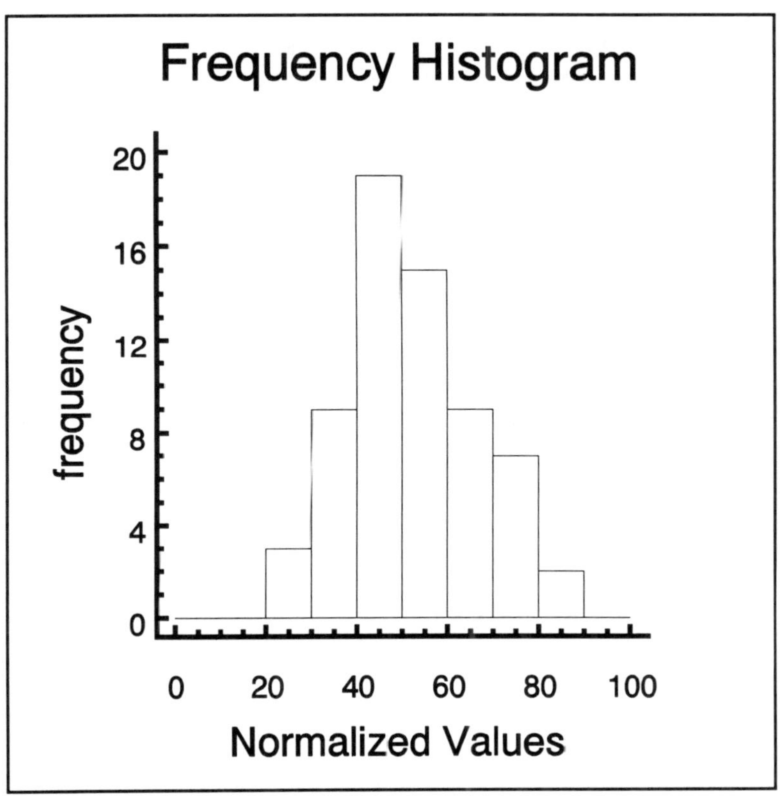

	Average
Average	52.2
Standard Deviation	14.7
Distribution	Normal

The average of the subgroup averages is 52.2 with a standard deviation of 14.7. The histogram depicting the distribution of the subgroup averages shows a very well formed normal distribution.

As a special note, the average of the individual values (our original random numbers) was 52 and the average of the subgroup averages is also 52. The fact that both sets of numbers results in the same answer gives us a tool to tactically detect a change.

Also the standard deviation of the individuals (STD_{Ind}) is 30 and the standard deviation of the averages (STD_{Ave}) is 14.7 STD_{Ind} is always bigger than STD_{Ave}. The technical name for the standard deviation of the averages (STD_{Ave}) is **standard error**.

The size of *n* depends on the original distribution and how much averaging is required to force the original distribution to normality. As the number of values grouped together increases, the closer to a normal distribution the grouped averages become.

What can the normalized numbers do for us? Since both the average of the individuals and the average of the averages are the same, we can use the subgroup distribution with its monitors of central tendency and variability to track and detect any changes in the true or individual distribution's center point. As the individual distribution shifts, so does the distribution of the subgroups. Both distributions will track or move in the same direction.

By tracking in the same direction, the distribution of the subgroups provides an indirect means of detecting a shift in the individual distribution. The advantage of using the subgroup distribution to detect change is that the subgroup's distribution is a normal distribution. This normality is a fact of nature and a guaranteed benefit. The assurance of the normal eliminates the need for testing the distribution. The subgroup assurance of normality allows us to always use the probabilities of the normal in our change detection system.

This detection of change is only a tactical tool to give quick and easy feedback of change. It **cannot** provide any strategic impact. This distinction is critical to assure that the correct set of books are always being used for the particular situation that is under analysis.

Why Use Subgroup Samples of 5? Since the increasing number of values in the subgroup simply normalizes the data, there is a point of diminishing returns, where more values clustered together does not improve the subgroup or average distributions normality. As a general rule of thumb, samples of five in a subgroup will normalize most distributions. In our example the flat distribution (nowhere near a normal) becomes normalized when five values are clustered together.

More values clustered together forming a larger subgroup will simply tighten the distribution about the mean. The subgroup size is strictly a means of normalizing our data and not a tool for assuring the accuracy of our analysis.

CAUTION ! Never try to use the tactical tools or numbers for strategic purposes. Strategic management reports that use customer specifications must use the individual readings to assess the impact to the customer. The use of subgroups and their averages has completely eliminated the ability to assess impact to a customer's specification.

Note the first subgroup that we analyzed. The actual five values in the subgroup from the table are 31, 94, 7, 0, and 1. The average of these values is shown below:

$$
\begin{aligned}
Average &= \frac{31+94+7+0+1}{5} \\
&= 26.6
\end{aligned}
$$

The average of 26.6 is a radical departure from the high value of 94 and low value of 0 in the group. To assess how well the values meet a customer's specification, we must look at the individual values.

Remember!! Analysis of the individual readings is used for strategic and management reporting. Analysis of the averages is used for tactical and operational reporting. Standard deviation is used for strategic reports while standard error is used for tactical reports.

Another advantage of using the analysis of the averages is that we have eliminated a decision of assessing the normality of the distribution. Our tactical or operational workforce may not have the statistical expertise that would be required if we only used the original or individual distribution.

What if the Original Distribution is Normal?

The figure below shows the distribution of the individuals that is normal. The subgroup average distribution is also shown in the same chart, and this distribution is much tighter around the center point.

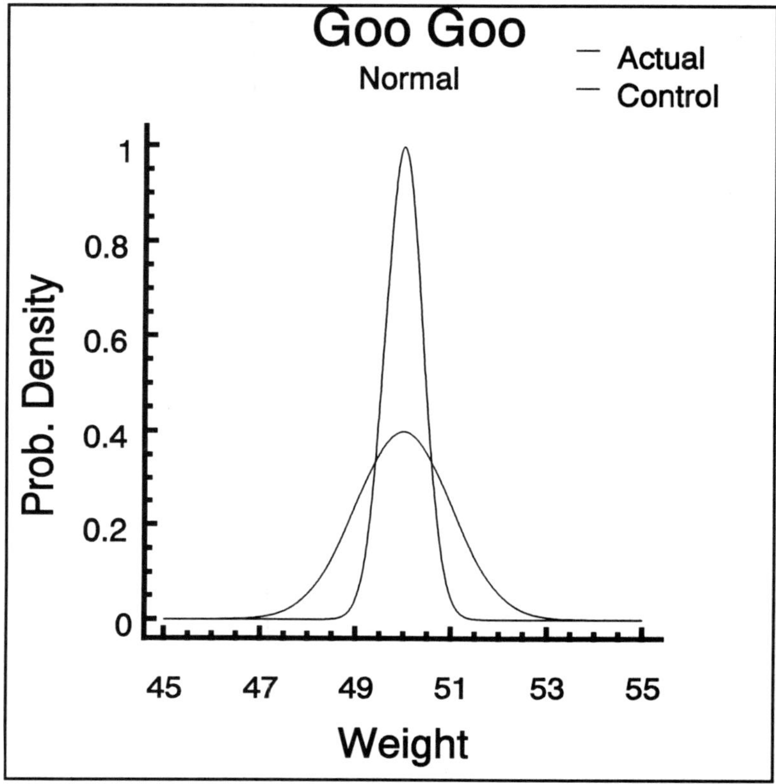

The chart above shows the result of normalizing an already normal distribution. The subgroup distribution becomes an even better normal distribution and also the distribution of the subgroup is simply a much tighter band around the center.

Central Limit Theorem

This process is called **central limit theorem**. Central limit theorem states that for any distribution, when samples of values are averaged together, the resultant distribution of the averages will be approximately normal. The larger the number of values averaged together, the closer to a normal distribution. This theorem works no matter how far away the original distribution is from a normal distribution.

Relationship of Standard Error and Standard Deviation

Standard deviation of the individuals is the monitor of the variability of the actual distribution. Standard error or the standard deviation of the averages is the monitor of the variability of the subgroup averages. Standard deviation is greater than the standard error. In mathematics this would be written as follows:

$$STD_{Ind} > STD_{Ave}$$

Since standard error is always smaller than standard deviation, a factor that would allow the two to approximately equal each other would be a tremendous tool. Nature has provided the factor, so let's investigate. The equation below shows the revised equation with an equality.

$$\frac{STD_{Ind}}{Factor} = STD_{Ave}$$

Standard error is approximately equal to standard deviation divided by the square root of the number of samples in the subgroup. The factor that must be replaced in our equation is the square root of the samples size. This sample size is the size of the subgroup that we have averaged. The most appropriate name for this factor is the **subgroup size**.

$$\frac{STD_{Ind}}{\sqrt{n}} = STD_{Ave}$$

For our example, the subgroup size is five for each subgroup. The standard deviation of our data is 30 and the standard deviation of the subgroup averages is 14.7 When these values are inserted in the equation, the following results.

$$\frac{30}{\sqrt{5}} = 13.6$$

The square root of five is roughly 2.2.

$$\frac{30}{2.2} = 13.6$$

The standard deviation of 30 divided by the square root of a subgroup size of five is 13.6 and happens in this case to be a little under the calculated standard deviation of the averages. The knowledge of the relationship of standard error and standard deviation allows us to build two sets of formulas for intervals.

The more samples that are analyzed, the stronger the relationship between the standard error and standard deviation will be.

Strategic Support

In Chapter One, we discussed the use of average, standard deviation, and the Z-value to build an interval. This basic idea of intervals will be used for two sets of analytical books to support Division of Labor.

Strategic decisions and goals

Strategic The art of devising or employing plans or stratagems toward a goal.

Management decisions are strategic and assess impact. Management or strategy selects, assesses, and provides the means. All strategic analysis must support this mission.

In a manufacturing quality setting, strategic issues deal with assuring that we have the correct facility to make what the customer wants. Thus we must compare our data to the customer's specification. To accomplish the strategic mission, the individual values must be used. To make the objective a more general statement, the strategist is responsible for providing the correct facility and for assuring that if the facility is changed in any way, it will still be capable.

The statistical tool using the individual values is called a **prediction interval**. Prediction intervals are based on individuals and the use of standard deviation. All the statistical tools based on prediction intervals will be the strategist's tool kit. The formula for prediction intervals is:

Prediction Intervals

$$X_i = X_{Ave} \pm z * Std_{Ind}$$

Tactical Support

Tactical decisions and goals

Tactical The art or skill of employing *available* means to accomplish an end.

Operational decisions are tactical and detect change. Operations or tactics use the available means. All tactical analysis must support this mission.

In a manufacturing setting, the operator is our tactician. He must use the available means to its maximum and optimum while he runs the facility as correctly and consistently as possible. Therefore, a separate set of tools which alert the tactician to change is required to support the tactician's mission of consistently running the available facility. The statistical tools called **confidence intervals** will be the tactician's tool kit.

Confidence intervals are based on averages and the use of standard error. All the statistical tools based on confidence intervals will be the tactician's tool kit. The formula for confidence intervals is:

Confidence Intervals

$$X_i = X_{Ave} \pm z * Std_{Ave}$$

We can show the relationship between standard deviation and standard error in the formula:

$$\frac{STD_{Ind}}{\sqrt{n}} = STD_{Ave}$$

Then we can substitute as follows:

$$X_i = X_{Ave} \pm z * \frac{Std_{Ind}}{\sqrt{n}}$$

What is the Analysis Support Team for Decision Making?

Tactical execution is required for strategic decisions to be effective. If operations fails to correctly run the facility, no strategy will work. Our strategic decision-maker must count on our tactical decision-makers to do their job, for only then can the strategic decisions be effective. Many strategic options are available, but they will work only if the operation can run smoothly.

To repeat, management can make good decisions only when operations is correctly and consistently running the facility. When this is the case, many options are available to management. When the plant is run incorrectly, we have no way of making a strategic decision that is based on sound logic.

The prediction intervals and confidence intervals become the eyes and ears for effective decisions. The correct set of books must be selected to provide the appropriate information to support the type of decision being made. The correct presentation of information is essential to good sound decisions. Failure to provide or choose the correct set of books and the correct analysis tool will yield incorrect information.

When strategic and tactical personnel both do their jobs, they form a team. Operations is in deep trouble when management does not do its decision-making job of providing the facility. If the facility (available means) is not capable, we are doomed to having dissatisfied customers and stockholders. If operations does not perform its role of running the facility (available means) correctly and consistently, then how can management assess the impact to the customer?

The correct selection of the analytical tool then becomes of paramount importance such that the tactician and the strategist can become a team. Just like strategic and tactical decisions form a team, confidence and prediction intervals form an essential team for properly prepared information. Both the decision-makers and the proper support information must come together to build a finely honed team that allows for good, effective, clear decisions.

When operations (tactician) and management (strategist) execute their jobs brilliantly, we can truly improve all three components of quality: performance, cost, and timeliness. Let's see an example of how tacticians and strategists should perform.

What are the Appropriate Decision-Making Tools for this Team?

There are many tools to support both tactical and strategic decisions. Therefore, we must match the correct situation with the correct analytical tool. A few of the analytical tools are listed below.

Operational Tools for Tactical Decisions:

- Control Charts
- Walkabout™s
- Orthogonal Arrays
- Planners
- Pert Charts
- Gantt Charts
- Resource Charts
- Mean Plots

Management, Customer, and Sales Tools for Strategic Decisions:

- Capability Studies
- Feasibility Studies
- Box Plots
- Resourcing

Continuous Operation We must understand Division of Labor in order to determine which analytical tools to use. The purpose of this book is to provide insight into the proper tools to run any continuous operation. The set of tools that must be used is called Statistical Process Control (SPC). This tool kit consists of a tactical tool (for operations) called control charts and a strategic tool (for management) called a capability study.

The control chart uses the set of books based on subgroup averages and the concept called confidence intervals. Control charts, which are based on confidence intervals, are operations' tool to detect change in the process. When a change is detected, the process is said to be "out of control." "Out of control" simply means something has changed.

The capability study uses the set of books based on individuals and the concept called prediction intervals. Capability studies, which are based on prediction intervals, are management's tool to assess any detected change in the process. When any change is detected in the process, management must assess the impact of the product to the customer's specification. Keep in mind we are assessing change, and change can be either an improvement or a deterioration. Thus both tools are essential, and nothing progresses if one is missing.

Chapter Summary

In the first chapter, we discussed using good technique based on the concept of point estimates. The three areas that must be monitored are central tendency, variability, and distribution. Now we have added the concept of intervals and central limit theorem. Central limit theorem allows us to build specialized tools for monitoring both strategic and tactical situations.

Normal Distribution — Probability

Central Limit Theorem — Forces Normality

$$\frac{STD_{IND}}{\sqrt{n}} = STD_{AVE}$$

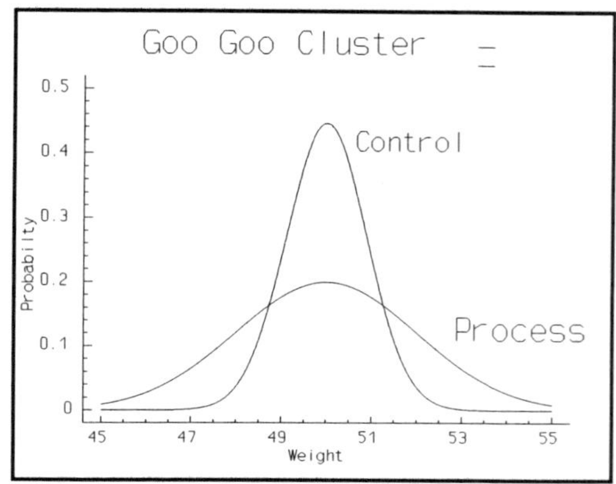

Statistical Support for the Second Principle Process Management

A clear understanding of Division of Labor is required to understand management and operational issues. Division of Labor is the driving force for all decisions and goals.

Second Principle of Process Management

Division of Labor is the framework for all aspects of decision making. It must be clearly understood to separate the strategic and tactical decisions. Operations makes the tactical decisions of running the facility. Management makes the strategic decisions of assessing the facility's suitability for the job.

Proper tools and tool selection are essential for the decision-making team and the analysis support team. Our understanding of Division of Labor must now be supported by proper analytical skills.

The following table summarizes the analytical tools to support tactical and strategic decision-making and goal-setting.

	Tactical	Strategic
	Operation ↓ Detect Change ↓ Averages	Management/Customers/Sales ↓ Assess Capability ↓ Individuals
Central Tendency	**Ave (Ave)** =	**Ave (Ind)**
Variability	$\mathbf{STD}_{Ave}$	$\mathbf{STD}_{Ind}$
	$\mathbf{STD}_{Ave} = \frac{\mathbf{STD}_{IND}}{\sqrt{\mathbf{n}}}$ Where n = subgroup size	
Distribution	Central Limit Theorem Normalizes Data	Histogram
	Confidence Intervals $X_i = X_{Ave} \pm z * \frac{Std_{Ind}}{\sqrt{n}}$	Prediction Intervals $X_i = X_{Ave} \pm z * Std_{Ind}$

All reports must assure that the proper set of books matches the proper decision type - either strategic or tactical. No one analysis tool can support both types. Both sets of books must complement each other and work as a team.

Now we are ready to examine the reporting flaw of improper grouping in Chapter Three.

Chapter Three: Proper Grouping

We learned in Chapter One that two major analytical flaws stand in the way of providing proper information. The two flaws are 1) technique and 2) grouping. For proper numerical analytical technique, we must monitor each of the following:

- Central Tendency
- Variability
- Distribution

The focus of Chapter One was to clarify the first flaw of improper technique. Failure to monitor all three technical characteristics will lead to catastrophic mistakes.

Now our attention must move to the second analysis flaw: **Improper Grouping**.

Flaw 2: The Grouping Flaw

In a *USA Today* article dated November 2, 1994, the state of airline commuter aircraft safety was discussed. Graphs, charts, tables of statistics, and the body of the article focused on the unsafe nature of the commuter aircraft. This quote appeared buried deep in the body of the text: "But the FAA counts any aircraft with one to 30 seats as a commuter, including single-engine planes, bush planes flying in Alaska - even helicopters.
In 1991 and 1992, 22 of the 45 commuter accidents were in Alaska or dangerous off-shore locations."

The purpose of the article was to focus attention on the poor safety of commuters. However, by improper grouping, both the writer and the government have received bad information that is causing major changes in airline regulations. Are the airline regulation changes necessary? In order to answer the question, proper grouping is essential. Using the reports as published in the article would lead to a flawed base for any decision. Thus any decision that is made based on this report is flawed.

One of the most glaring problems with statistics and analysis is the improper grouping of our data. Just because we are using proper statistical technique does not mean that we are receiving accurate information. Grouping the ten machines below to report a 1% defect rate paints an incorrect picture of the overall performance. No machine is actually running at a one percent defect rate. Not only are none of the machines running at a one percent defect rate, no machine is even close to the one percent.

Machine	#1	#2	#3	#4	#5	#6	#7	#8	#9	#10
% Defective or Scrap	0%	0%	0%	0%	0%	0%	0%	0%	0%	10%

The following are quotes from two of the masters of statistics--W. Edwards Deming and Eugene L. Grant. Their statements are clear as to the importance of proper grouping and leave no room for ambiguity.

In the book ***Out of the Crisis***, Deming says in his section called **A first lesson in application of statistical theory,** "Courses in statistics often commence with study of distributions and comparison of distributions. Students are not warned in classes nor in the books that for analytic purposes (such as to improve a process), distributions and calculations of mean, mode, standard deviation, chi-square, t-test, etc. serve no useful purpose for improvement of a process unless the data were produced in a state of statistical control."[1]

In ***Statistical Quality Control***, Grant says, "The statistical tests discussed in this book begin with the presumption that samples are being drawn from a single universe. Until the control charts indicate that the presumption is true, estimates of the parameters of the distribution underlying the process are virtually meaningless. All the pitfalls regarding rational subgrouping apply, with special emphasis to capability studies."[2]

There are three types of groups we must consider 1) Physical, 2) Associational, and 3) Time-Series. We will discuss all three in the next section.

Group Definition In analysis and reporting, we must make certain assumptions and definitions. Therefore, all users must clearly understand the definition of the terms. A key assumption (both Deming and Grant understood and stated the importance of this assumption) is that all analysis is based on groups of like and homogeneous kind. In this context, the term homogeneous has the following requirements:

1. **Physical Groups** are made of things that are materially the same. Members of physical groups are things having like kind such as type, size, color, same machine, or operator.
2. **Associational Groups** are made of things that are similar but not necessarily exactly the same. Their members are also of like kind.
3. **Time-Series Groups** are composed of material that is similar over time. Members of a time-series group are manufactured at the time when consistency is demonstrated. This consistency is called *in-control.*

The *why* of homogeneous is critical. To more fully understand accurate analysis and reporting, the following example and scenarios will be used. A key ingredient is the assurance of homogeneous groups. The examples are to assure we can visualize and understand the concept.

---

[1]***Out of the Crisis***; *W. Edwards Deming*. Page 312.

[2]***Statistical Quality Control*** - Sixth Edition; *Eugene L. Grant and Richard S. Leavenworth*. Page 167.

Physical Groups - Certain aspects of the word *homogeneous* are obvious. We cannot mix size 12 shoes with size 9 shoes in a homogeneous group. Neither can we mix blue shirts with white shirts because an average of a 10½ size shoe or a pale blue shirt is misleading. Instead, simple common sense is required to properly prepare physical groups. Physical groups appear to be obvious to keep separate from one another, but as we saw from the excerpt in the *USA Today* article, this is not the case. Everyday articles, radio and television newscasts, and daily conversation all make the grouping mistake.

Other grouping requirements are even more difficult to conceptualize and even harder to assure. To pass the many difficult requirements of a homogeneous group, we need statistical tools like Statistical Process Control (SPC) for a time-series check and Analysis of Variance (ANOVA) for an associational check.

Associational Grouping - Associational groups are things of like statistical kind. The machine that runs at 10% defect versus a machine that runs at 0% defect are associationally different. Analysis of Variance (ANOVA) is a tool that tests for grouping by association. Association is another way of saying that we only group those thing of like kind. These groups are then called *homogeneous*. Non-homogeneous items must never be placed in the same group for analytical reports.

Time-Series Group - Time-series groups are things that are similar over time. The machine that produces 10% defective product for one hour out of twenty-four would require separate analysis of the one hour at 10% versus the other 23 hours at 0%. The control chart function of SPC provides the vehicle to assure a state of statistical control and validate the time-oriented grouping. Statistical Process Control is another way of saying *homogeneous* or *consistent over time.*

The term homogeneous has the following requirements:

1. Product having like kind, type, size, color, characteristics, etc.
2. Product manufactured by the same machine and same operator.
3. Product manufactured at the same time.
4. Product manufactured using material from a common source.
5. Product does not have to be exactly the same.

Homogeneous groups may seem obvious to one person and not to another. Thus clear guidelines are required.

These other requirements are very difficult to conceptualize and even harder to assure. To pass the many difficult requirements of a homogeneous group, we need a statistical tool like SPC (Statistical Process Control). The control charts' function of SPC provides the vehicle to assure a state of statistical control. Statistical control is another way of saying homogeneous or consistent. We must never allow the mixture of non-homogeneous product groups.

Why Homogeneous?: The why of homogeneous is critical. The following examples are designed to show us the advantages of correct group formations and the pitfalls of incorrect group formation. A key ingredient is the assurance of homogeneous products when grouping. The examples are exaggerated to assure we can visualize and understand the concept.

Candy Example

As an example for our proper grouping, we will use a facility that makes candy. Our peanut product must fall inside a weight range of 45 grams to 55 grams to meet our customer's requirement. Any candy cluster that weighs less than the 45 grams is classified as a defect.

Candy Grouping Examples - We must never allow the mixture of non-homogeneous physical, associational, or time-series groups because this will create erroneous reporting. The following case study gives examples that will explain these problems. We will visit our candy customers and manufacturing facility to explain some of the problems with improper groups.

Physical Grouping - Two types of candy are made in our facility. One contains peanuts while the other contains pecans. For proper analysis, the pecan and peanut product should not be mixed into one group. In addition, the peanuts are more dense than the pecans, thereby making the peanut product weigh about 50 grams while the pecan product weighs roughly 42 grams. Grouping an equal number of both products would yield an overall weight of 46 grams. The overall weight of 46 grams means absolutely nothing. Our customers always specify the specific product that they are interested in procuring.

Customer Order Background - We have a very demanding customer for our candies. He has just placed an order for 10,000 good peanut candy clusters. The customer wants 1000 candy clusters shipped in each of 10 containers. This customer is willing to give us some room on performance since he understands the failings of humans. He has established that he will tolerate a small amount of defects. *This customer expects and wants no defects*, but he will not penalize us if we accidentally have no more than 500 defective units in his order. As stated in the contract, the defect rate cannot exceed 5%.

Our customer has also devised a method to assure enforcement of his requirements. Since we would destroy his product if we attempted to test all 10,000 clusters, he has devised a testing program which requires us to test 500 unbiased parts from each group of the order. The term for unbiased is *random*. If, of the 500 test products, more that 50 defective products are identified, then the order will be cancelled and all 10,000 parts returned.

Manufacturing Background - Our candy plant has 10 identical machines working on the peanut candy clusters. Each machine today is making the peanut candy product. This product is the same product the customer just ordered. At the end of the shift, each machine has manufactured 1000 units of clusters. The scrap rate for each machine is listed in the table below.

By reviewing data at a summary level, we run the risk of clouding or masking our problems. Take special note of machine 10 and the scrap it produces.

	Mach 1	Mach 2	Mach 3	Mach 4	Mach 5	Mach 6	Mach 7	Mach 8	Mach 9	Mach 10	Tot.
% Defective or Scrap	0%	0%	0%	0%	0%	0%	0%	0%	0%	100%	
Product Produced	1000	1000	1000	1000	1000	1000	1000	1000	1000	1000	10000
Scrap	0	0	0	0	0	0	0	0	0	1000	1000

To calculate a summary plant percent, we divide the total scrap by the total production.

$$\text{Plant \% Defective} = \left[\frac{\textit{Total Scrap}}{\textit{Total Production}}\right] 100\%$$

$$= \left[\frac{1000 \textit{ parts}}{10{,}000 \textit{ parts}}\right] 100\% = 10\%$$

Our plant percent defective is 10% in the summary. By grouping these together, and simply looking at the summary information, we have the picture that we are running a fairly efficient operation.

The product coming from each machine is put into a container holding the 1000 parts from only one particular machine. Therefore, the containers for product from machines 1 - 9 each contain 1000 good parts, but the container for machine 10 contains 1000 bad parts.

Grouping For Shipping - For shipping, all the containers are grouped into one group. We have incorrectly called this a quality assurance group instead of a shipment. Let us discuss the ramifications and consequences of this incorrect definition.

We have now shipped the 10,000 parts to the customer. The customer's order specified that 500 candies would be tested to assure that the candies were good. To avoid a biased test, the customer develops a testing plan to test equal quantities from each container of the total 500 test parts. The 500 piece test plan from the containers is shown below:

Container Machine No.	1	2	3	4	5	6	7	8	9	10
# of parts Sampled	50	50	50	50	50	50	50	50	50	50

The customer pulls his test values from each container. Since only good product is in containers 1 - 9, we will obviously have only good test results. But in container 10, produced on machine 10, all the product is bad, so the test samples must all be bad. The test results are shown below:

Container Machine No.	1	2	3	4	5	6	7	8	9	10
# of parts Sampled	50	50	50	50	50	50	50	50	50	50
Defective Parts	0	0	0	0	0	0	0	0	0	50

By incorrectly identifying the group, we have now put ourselves in the position of losing the complete order. No one should be surprised that the complete shipment of 10,000 candies is returned. All the good product is being lost because the bad product from the one bad machine is mixed in the shipment.

If we had tested true groups, we would have scrapped all the product from machine 10. In doing this, we could ship our customer 9000 good parts. By correctly forming the group, we could increase our productivity or throughput because no product is returned. In other words, the 10,000 parts that include the scrap are returned, while the 9000 parts would be accepted. Generally, though, we do not have such an obvious situation as this illustration. The next example is a more complex situation.

Mixing of Product - Imagine the same production schedule with an additional step to optimize the production system. This process allows us to thoroughly mix all machines' goods in a vat to form one large homogeneous group such that no product can be identified with its producing machine. This group is now a *true group*. This chart shows the **original** production and product shipment containers.

Container Machine No.	1	2	3	4	5	6	7	8	9	10	
Product Produced	1000	1000	1000	1000	1000	1000	1000	1000	1000	1000	10,000
Scrap	0	0	0	0	0	0	0	0	0	1000	1000
% Defective	0%	0%	0%	0%	0%	0%	0%	0%	0%	100%	10%

The problem with the mixed group is that we have contaminated good product with bad. Now the containers will have product that is like the example in the table below. As you can see, a few bad candies are in each container.

	CONTAINER										
	A	B	C	D	E	F	G	H	I	J	TOTAL
Product Shipped	1000	1000	1000	1000	1000	1000	1000	1000	1000	1000	10,000
Good Parts	900	900	900	900	900	900	900	900	900	900	9000
Bad Parts	100	100	100	100	100	100	100	100	100	100	1000

The odds are very high that we will still lose the entire order in testing. However, the dilemma really occurs if we don't lose the group in testing, and we ship this order. *We have shipped defective goods to the customer, and we should not be surprised to find the customer unhappy.* We have shipped more bad product than the customer's order had stated would be acceptable. Remember, the customer stated that he wanted no bad parts at all and would tolerate no more than 500 bad parts in the order.

As is now obvious, proper group formation is vital to accurate reporting. Each machine's product is kept separate, and each homogeneous group can stand on its own. Once a group has been identified at the detailed level, these numbers can then be grouped into an aggregate report. Accurate and effective decisions can be made about each group when we avoid any contamination. By correctly forming groups, we are able to increase our throughput because we will not allow any bad product contamination.

Time-Series Group - A more complex and less obvious example is a time-series production problem. We are working in the same plant but analyzing only one machine. Today this machine is making the customer's ordered product. Every hour of the shift the machine manufactures 1000 candies. The scrap rate for each hour is listed in the table below. Since a container holds 1000 parts, each hour's product is put into its own container.

Examples	PRODUCTION TABLE									
	1:00	2:00	3:00	4:00	5:00	6:00	7:00	8:00	9:00	10:00
Parts Produced	1000	1000	1000	1000	1000	1000	1000	1000	1000	1000
Scrap Parts	0	0	0	0	1000	0	0	0	0	0
% Defective or Scrap	0%	0%	0%	0%	100%	0%	0%	0%	0%	0%

All the machine's product from 1:00 through 10:00 is grouped together into one shipment as the product is produced. We are incorrectly calling this total production (all 10 hours and 10,000 units) a group. We are taking an extremely big risk of rejecting or accepting the complete shipment.

We are filling the containers each hour as the goods are produced. The product in each container is shown below.

	CONTAINER									
	1:00	2:00	3:00	4:00	5:00	6:00	7:00	8:00	9:00	10:00
Container Number	A	B	C	D	E	F	G	H	I	J
Parts Produced	1000	1000	1000	1000	1000	1000	1000	1000	1000	1000
Bad Parts	0	0	0	0	1000	0	0	0	0	0
Good Parts	1000	1000	1000	1000	0	1000	1000	1000	1000	1000

The testing plan used by the customer will be similar to the prior plan and is shown below.

Container Hour	1:00	2:00	3:00	4:00	5:00	6:00	7:00	8:00	9:00	10:00
Container Number	A	B	C	D	E	F	G	H	I	J
# of parts Sampled	50	50	50	50	50	50	50	50	50	50

Once again, we see how the group will fail. Hours 1:00 - 4:00 and 6:00 - 10:00 have produced only good parts while the container for the 5:00 hour has only bad parts. We see in the table below that the 5:00 container sample will have 50 bad parts. We once again should not be surprised to find that the customer will be unhappy, and the order will be returned and cancelled.

Container Machine No.	1:00 A	2:00 B	3:00 C	4:00 D	5:00 E	6:00 F	7:00 G	8:00 H	9:00 I	10:00 J	Total
# of parts Sampled	50	50	50	50	50	50	50	50	50	50	500
Defective Parts	0	0	0	0	50	0	0	0	0	0	50

Production must be defined as only good product that the customer will accept. For us to maximize production, we must never allow bad parts to contaminate good parts. To maximize production, we must assure that no defective parts are knowingly mixed in good parts through combining of machines or hours. Once we have assured that all groups (i.e. machines, hours, etc.) are the same, then and only then, can we begin to mix.

Mixing of Hours - Imagine the same production schedule for one machine, with an additional step to optimize the production system. This process allows us to mix all the machine's goods for many hours into one vat. Each hour's product is mixed together to form one large homogeneous group called a *true group*.

The machine's product is mixed together to form a mixture. This group is now a true group but has obvious problems because we have contaminated good product with bad. The odds are very high that we will lose the order in testing. However, again the real dilemma is that if we don't lose the group in testing and we ship this order, *we have shipped defective goods to the customer. We should not be surprised to find the customer unhappy.*

Improper Associational Groups We may never have a situation as obvious as the prior ones but imagine the following example for a machine. Each machine's production is lumped together forming inappropriate groups. Even when the differences are not so dramatic, the impact of bad reports is the same.

Machine	1	2	3	4	5	Total
Parts Produced	1000	1000	1000	1000	1000	5000
Parts Scrap	2	3	2	1	50	58
% Scrap	.2%	.3%	.2%	.1%	5%	1.2%

By mixing machines 1-4 with machine 5, we have dramatically affected the defective percentage for machine 1 through 4 in a very detrimental way. We have the potential of contaminating good parts from machine 1-4 with the bad parts from machine 5. An analysis of this incorrectly grouped example will generate a completely false impression. The 1.2% scrap rate represents absolutely nothing.

Chapter Two showed us the importance of intervals. Intervals are an effective and appropriate mechanism to make associational groupings. The chart below gives an example of how the groups can be formed.

The graph to the right is called a **mean plot**. The mean plot is one of the tools for discrete group comparisons. The chart shows that product 1 is distinctly different from products 2 and 3. These three products are the same physically, but the weights are different to make two separate associational groups. One group contains products 1 and 2 while the second group contains only product 3.

Group Comparison through *Mean Group*

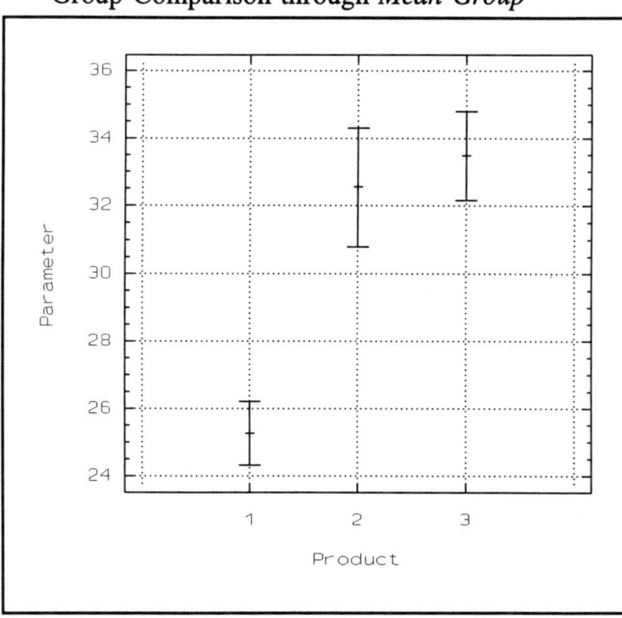

Improper Time-Series Groups Imagine the time-oriented production for a machine as shown below. Each hour is lumped together with no thought to proper grouping.

Time	1:00	2:00	3:00	4:00	5:00	6:00	7:00	8:00	Total
Parts Produced	1000	1000	1000	1000	1000	1000	1000	1000	8000
Parts Scrap	2	2	100	100	0	1	1	2	208
% Scrap	.2%	.2%	10%	10%	0%	.1	.1%	.2%	1.2%

The pocket of bad product produced from 3:00-4:00, if allowed to be mixed with that of the other hours, will contaminate the product of all other hours. A report stating that this machine ran a 1.2% scrap rate is just as incorrect as the improperly grouped machines.

We would not be surprised if the customer and management complained about these mixed and contaminated goods.

The SPC tool called control charts is the tool to monitor any shift in the time-series groups. Any shifts from the usual running characteristics of the machine must be investigated.

The chart to the right shows the weight of peanut candy clusters when they are produced. The weight of the clusters has shifted from time period A to time period B. A second shift is shown from time period B to time period C.

Candy Weight by Time of Production

A + B + C = 1%
A = 0%
B = 6%
C = 0%

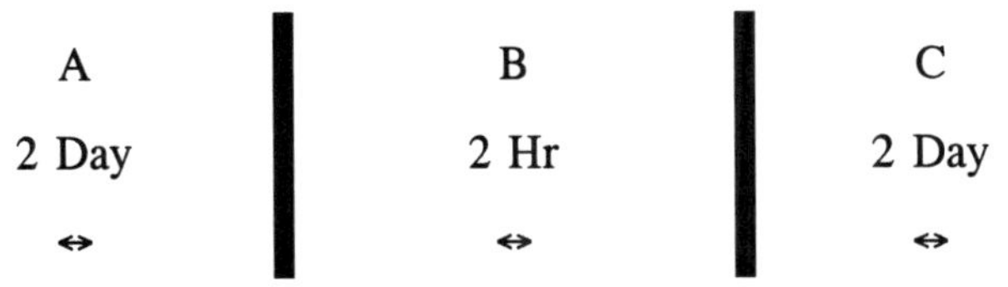

A	B	C
2 Day	2 Hr	2 Day
↔	↔	↔

The defective percentage for time period A is zero, for time period B is six, and time period C is zero. Two distinctly different groups can be seen. If we attempted to group all the time periods together (A, B, and C), we would compute a one percent defect rate which would be completely misleading.

Proper Group Formation Summary

We can change the scenario by adjusting the percentages and the way in which the groups are formed, but correct formation of a group is critical to any report or analysis.

The product and process data must be kept separate by grouping so that we can identify homogeneous groups and be able to form an aggregate group.

In summary, we must group based on:

- Physical similarity
- Associational characteristics
- Time of manufacture

We can assure proper physical groups by employing common sense. The associational groups are identified by use of a technique called Analysis of Variance. The time-series groups are formed based on the analytical tool called control charts.

With proper group formation we can expect the following benefits:

1. Avoiding the contamination of good product with bad product.
2. Never failing to take action on bad product or processes to remove its cause.
3. Never failing to notice an improvement so that we can continue the cause.
4. Increased product throughput.
5. Increased production and productivity.
6. Improved process.

...and the list of benefits just continues.

Data Collection: Our data collection objective is to provide information to support better decision-making. Each piece of data should have the clear purpose of supporting a specific decision, and this purpose should be clearly stated for all who are gathering the data. By having a clear purpose as the environment changes, the data gatherers can adjust the manner in which they collect the data. Should the specific decision no longer be needed, then the data collection can also be discontinued.

Management needs information to make intelligent decisions. This information must be prepared in a format that concisely and accurately presents factual data. The problem with a detailed analysis is that one can become overwhelmed with data. We can "lose the forest for the trees" if we don't condense. The trick is to condense without making the "grouping" mistake. An alternate plan is to aggregate (summarizing of like kind) the data to carefully and logically create a series of homogeneous groups for summary. Exceptional groups may have only one member!

For us to do the detailed analysis that will lead to accurate reporting, a great deal of data gathering will be required. To adequately handle the detailed reporting, all associates must participate in the data gathering effort. To support proper group formation, a bottom up strategy of data reporting is required. This bottom up group formation will test for both time-series and associational groups.

Bottom Up Strategy

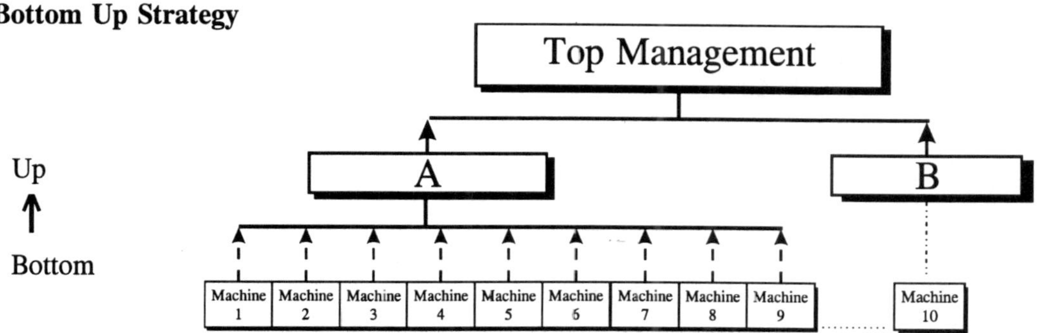

This bottom up strategy will document clearly the status of the area. This proper grouping will allow correct and accurate information to flow to top management. With better information in hand better decision-making is expected.

Capability Requirements from SPC

Each machine's product must be kept separate by time frame such that we can identify homogeneous groups. The only alternative to assuring correct formation is to assure the following:

1. Make sure each individual machine is monitored.

2. Use the SPC tool called control charts to identify when the time-series homogenous group has shifted.

3. Make sure the monitoring is set at an appropriate time interval for each machine to assure a continuous surveillance. The time interval must be set such that any time-series shift can be detected.

4. Make sure a detailed investigation is undertaken when a shift has been noted. This investigation must determine the cause. The effect of the cause can be either an improvement or a deterioration.

5. Take action to correct any deteriorating causes and remove any contaminated product.

6. Take action to continue the cause of any improvements.

Chapter Summary

For any analysis to be on solid ground, the following must be the center piece to any study.

Technique

For proper numerical analysis, our technique must monitor each of the following:

- Central Tendency
- Variability
- Distribution

Failure to monitor any one of the three will have dire consequences.

Grouping

Our groups must form one universe (homogeneity) based on:

- Physical similarity
- Associational characteristics
- Time of manufacture

We can assure proper physical groups by employing common sense. The associational groups are identified by using a technique called Analysis of Variance. The time-series groups are formed based on the analytical tool called control charts.

Proper Analysis

All our analyses are based on the assurance that both proper technique and groupings have been used. This combination will allow us to provide good information. This analysis will require a bottom up strategy of data collection and analysis.

Chapter Four: Statistical Process Control

Statistical Process Control (SPC) is used to analyze and monitor a continuous process. SPC should be used to monitor both product and process characteristics of all continuous operations.

Since SPC is designed to monitor a continuous process, an ongoing surveillance program is required. The flow to the right depicts the steps required for this monitoring system. First, we must gather data on a regular and consistent basis. Second, this data is measured for consistency with older data using a control chart. Once a state of statistical control is established, we must begin the third step of assessing how well the data meets the customer's needs using a tool called a capability study. The final step is the maintaining of control with a capable process.

Each one of these steps will be explained in this chapter and then a following chapter will discuss all the details associated with each.

Statistical Process Control is comprised of the two following tools 1) control charts and 2) capability studies.

A **Control Chart** is designed as an alarm system to detect change (both improvement and deterioration) in an operation. Because it is the operator's tool for monitoring a particular machine or system during its time of function, it includes an alarm system for detecting any type of shift or change. This alarm must be dynamically re-calibrated to reflect the current status. The following are types of control charts.

$\bar{X}$ & R, Individuals
P, C, U

A **capability study** is the tool for management to assess how well the product being produced matches the product the customer desires.

The SPC Team includes both control charts for tactical purposes and capability studies for strategic work. Therefore, both must work together to form a cohesive team. Let's first examine each tool independently. Then we will understand how the SPC team of control charts and capability studies works together.

Monitoring of a Continuous Process

Monitoring the Candy Plant Packaging

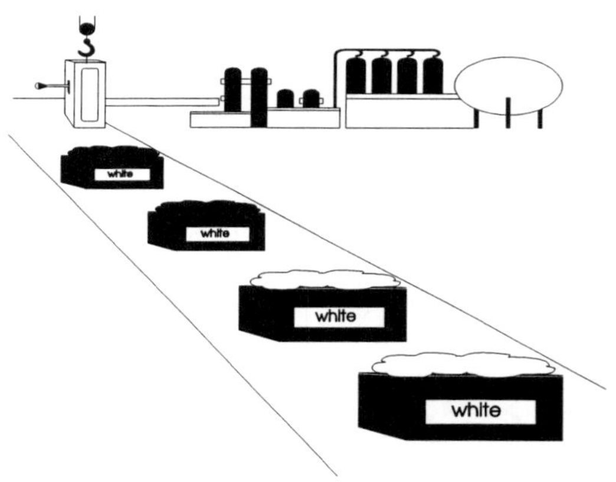

Notice in the drawing to the right that we are packaging our candy in a white container. The third product shown is not white but has changed to black, and all subsequent product is black.

Analysis of a particular machine, process, product, system, or entity must be done in the chronological sequence. If we lose the integrity of the time of an event, we have lost a tremendous source of data.

This knowledge of <u>when</u> the candy package changed is critical to our analysis. The quicker the detection of change, the better. The longer the package change goes undetected, the more the bad product builds. We must not wait until many bad candy packages are produced. If we had lost the time of the product change, we might have mistakenly determined that the change was a random event rather than a systematic change. Instead, we must have a continuous surveillance to detect change as early as possible.

Time Sequence In order to detect patterns, trends, and shifts, we must not only document the time of occurrence but also group the values in chronological sequence to detect the shifts. Chronological sequence is necessary to identifying causes.

If we used bands of data where the candy packages had changed (incorrectly) from white to black undetected, we would have an incorrect impression of the quality level of the product. The result would be very camouflaged and confused. No decision from this kind of analysis would be valid.

Time-series By maintaining the integrity of the time sequence, we also have the ability to take care of our time-series grouping. In Chapter Three, we saw the proper grouping as a major problem. The time-series grouping is resolved by using control charts.

The SPC objective is much more than simply detecting a defect. Rather, the SPC objective is to monitor for both improvements and deteriorations allowing us to get the most from our facilities. Like the United States army says, "Be all you can be." We want to use SPC to get all we can out of our facilities.

This objective uses the Second Principle of Process Management to make good decisions. The next section is a brief review of this principle.

Second Principle of Process Management Summarized: Division of Labor drives the decision making process. It dictates who will be in charge of the different kinds of decisions (strategic or tactical). We are doomed to a state of Quality Hell if the managers of the future fail to fully grasp Division of Labor.

Second Principle of Process Management

Division of Labor is the framework for all aspects of decision making. It must be clearly understood to separate the strategic and tactical decisions. Operations makes the tactical decisions of running the facility. Management makes the strategic decisions of assessing the facility's suitability for the job.

Control is a major tactical issue associated with operational issues. Capability is a major strategic issue associated with management, customer, and sales issues.

If we have a clear understanding of the Second Principle of Process Management, or this "Division of Labor", we will possess the key to appreciating the difference between tactical and strategic decisions.

Division of Labor requires a clear understanding of the differences between management/customer/sales issues and operational issues.

Tactics The art or skill of employing *available* means to accomplish an end.

Operational Decisions are tactical and detect change.

Strategy The art of devising or employing plans or stratagems toward a goal.

Management Decisions are strategic and assess impact.

Division of Labor drives and controls goal setting and decision making. Clarity of purpose is essential to effective goals and decisions. Operations must be tactically driven while management/sales/customer must be strategically driven.

The Principle of Division of Labor will have a profound an effect on decision making. The concepts below must be clearly understood, for without them, no improvement is possible.

Second Principle of Process Management
Division of Labor

↓

Accountability

	Operations	**Management/Sales/Customer**
Accountability	The correct and consistent running of *THE facility*.	Finding customers who will buy our products.
Responsibility	Running *THE provided facility* correctly and consistently	Providing *THE facility* that will produce what the customer wants
Missions	Control	Capability • Customer Specifications • Targets • Goals • Requirements
Decisions	Tactical	Strategic

For the most effective and efficient operations, we must understand our role according to Division of Labor and then execute brilliantly.

Strategy comes from the Greek word *strategos* which means "the art of the general." Generals make several kinds of strategic decisions including which resources to use, when to use the resources, and where to use the resources in the conduct of battle. Strategy, for our context, is the planning and providing of the correct resources at the correct time and at the correct place where they are needed.

Tactics is the *execution* of the strategic plan that the general has developed. We will expand on both tactical and strategic decisions in this chapter and in Chapter Six: *The Art of the Business Genius*.

Accountability: To be held accountable, we must first know whether we are in a tactical area or a strategic area. *Tactical personnel* should be held accountable for issues and decisions that deal with the execution of the plan. In a continuous operation, the plan deals with the facility that has been provided. The operator, who is the tactician, should be held accountable for the correct and consistent running of the facility.

Strategic personnel should be held accountable for developing a plan and building a facility to meet the company's needs.

For a complete and detailed explanation of Division of Labor, I suggest my most recent book *Become the Manager of the Future*.

Second principle violations must be understood and avoided in order to effectively make decisions. The kind of situations that we will encounter if we violate the second principle are:

Violation 1 - No Thought to Capability

If we violate the second principle by confusing the responsibilities of management with those of operations, we have no basis for assessing the capability of the facility. We in management can only ask operations to do what the facility is capable of doing. In this way we can hold operations accountable for the correct and consistent running of the facility. Management must provide operations with a facility that is capable of producing the customer's product.

Violation 2 - Forced Capability

Good Trooper Award: If we violate the second principle by trying to place the responsibility for strategic decisions on operations, we will not be able to meet the customer's demands, and we also make a more defective product than was necessary in the first place. ***The harder a good associate works with improper instructions (violating the second principle), the poorer the quality.*** Quality will continue to degenerate.

Violation 3 - Poor Productivity

If we violate the second principle on an extremely capable process, we will make this machine appear less capable than it really is, and we must conclude that we are not running this machine correctly and consistently. Instead, we must run the machine to its own personality as correctly and consistently as possible. We must not be lulled into accepting products which are only as good as the customer wants and allow the process to float inside the customer's specification.

We must guard against any violation of the second principle--Division of Labor. A good associate who is put in the situation of having to run a facility with improper directions that violate the second principle is doomed to disaster. The better and more conscientious the employee, the worse the situation will become because the person will go to any lengths to carry out those improper directions.

Because of Division of Labor, SPC is comprised of two tools specifically designed to support either tactical or strategic decisions. The tactical tool is called a control chart while the strategic tool is called a capability study. Let's now focus our attention on the operator's role with analytical support for him to best run his machine. This effort is referred to as **control**.

Where are we?? Before any improvements or problems can be solved, knowing where we are is essential. Statistical process control is the set of tools that provides a clear idea of where we are in relation to our target. Before an archer can hit the bull's-eye, he must know where the target is. He must also know where his arrows are falling so that any adjustments are made in the appropriate direction.

The details of archery are essential to hitting the bull's-eye. These same details are essential to a continuous operation. Now let's focus on the first detail of the correct and consistent running of the operation.

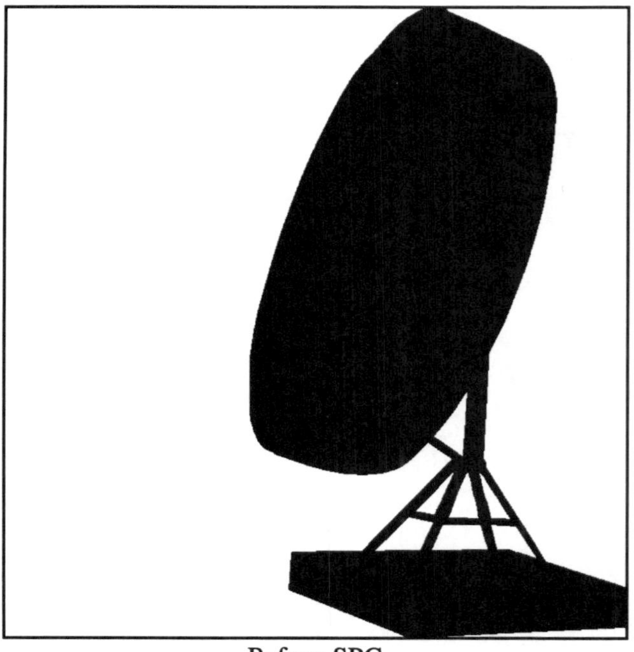

Before SPC
Where are we?

What does Control mean?? When we say that we are in control, we are only talking about operational issues. The term *in control* means that we are running the facility correctly and consistently. This gives us a way of knowing where we are.

We must have tools available to us to measure a facility's uniquely correct and consistent running parameters. Control charts were designed expressly to do the job of evaluating the correct and consistent running parameters of a machine. A control chart also functions as an alarm system to alert us to changes and deviations from these optimum parameters. Remember that changes can take the form of either improvements or deterioration of the process. However, control charts cannot identify whether the change is an improvement or a deterioration; *a control chart's mission is very specialized: to identify change only.*

Operational Decisions: When we understand the concept of control, we realize that the objective of operations is the correct and consistent running of the currently available and existing facility. Operations must keep a focus on the tactical needs by making tactical decisions and conducting tactical analysis.

Tactical decisions are defined as decisions employing available means to accomplish an end. Operations makes tactical decisions.

Operations => Tactics: Operations uses tactical decisions to continually better understand the facility. With this knowledge, we can use tactics to continue to correctly and consistently run the facility.

What is the Operator's Role?

The operator must focus his attention on the provided machine. The operator must run the machine as correctly and consistently as possible with the current state of knowledge. Running the machine to the maximum is the operator's responsibility for which he must be held accountable. Tools to assist the operator must be provided with the clear mission of supporting the correct and consistent running of the machine.

The objective of correctly and consistently running the machine means getting the most out of the machine no matter what the customer's specifications require. Our tools to support the operator will thus be independent of the customer and his requirements.

Which Tools to Support the Operation of a Continuous Process?

There are many tactical analytical tools that are designed specifically to support operations. One tool, called a **control chart**, is used on continuous operations to detect change. Control charts provide the mechanism to determine whether or not a particular machine is running correctly and consistently, and are also a means of monitoring and detecting homogeneous groups. They could also be called "consistency charts," "change detectors," or "cause detectors."

This is an example of a control chart:

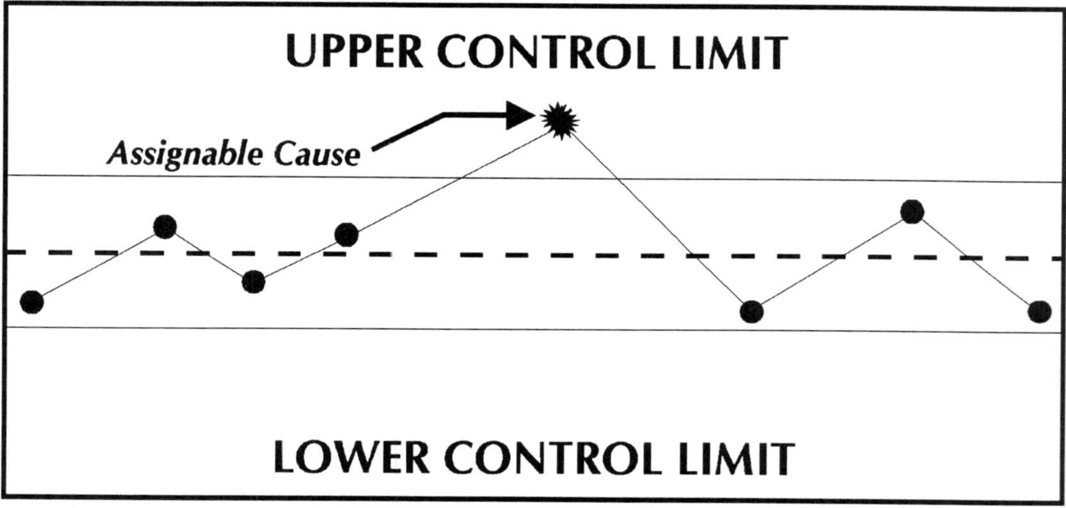

Many people attempt to make a control chart into both a management and an operations report by including specifications, goals, customer expectations, and tolerances on them. Control charts are operations tools, ***not management reports.*** Since specifications are strategic in nature, they NEVER appear on a control chart. The whole mathematical concept behind control charts assumes we are only discussing tactical issues. We will be concerned with specifications in the discussion of the strategic tool called "capability studies" discussed later in this section.

Control charts were <u>never</u> intended to be used as an indication of how well the product fits the customer's needs. Control charts are for the tactical decisions of correctly and consistently running the facility and for detecting change.

What Exactly is a Control Chart? The objective of a control chart is to help an operator run a machine correctly and consistently (without change). This is referred to as being *in control.* Because change can be either an improvement or a deterioration, any change must be identified immediately so that management can assess the impact to the customer.

Establish Process Control

Since the change must be detected quickly, a change detector is critical. That is exactly what a control chart is.

I hope you have a smoke detector in your house. The smoke detector sits quietly in the background until smoke or fire is detected. When the detector senses these situations, it sounds the call to arms loudly and aggressively. No action is taken based on the alarm alone, but immediate inquiry into the situation is required. We must know the cause of the fire before taking action because if we throw water on a grease fire, the fire will spread. This is exactly how a control chart works - instantly alerting the operator to any change.

When we are in control, we are running the machine in a steady fashion. *In control* is another way of saying that the machine is running with no change from the machine's typical running results.

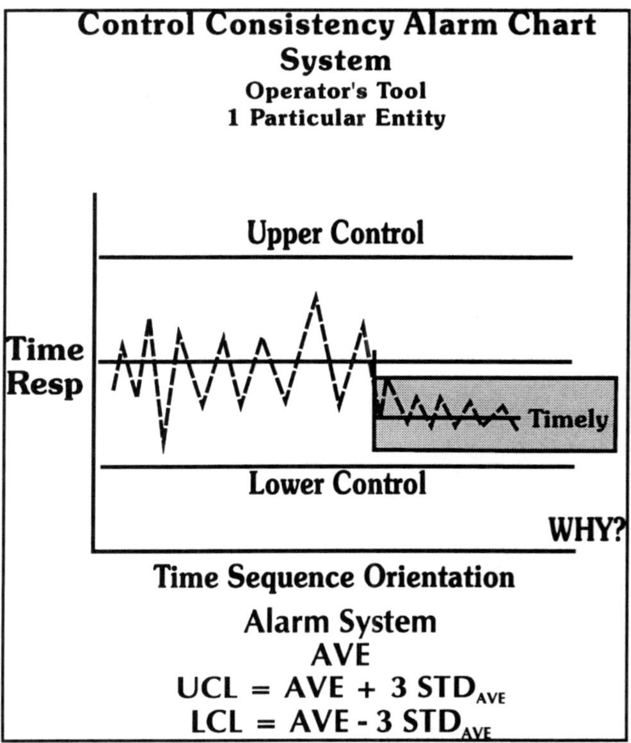

Why is Timely Reporting Essential?

In the early 70's, the comedian Flip Wilson had several characters that he portrayed in sketches. One of my favorite characters was Geraldine. She was well-known for her lines, "What you see is what you get," in reference to herself and, "The Devil made me do it," when she was in trouble. She attended the Reverend Leroy's church which was named **The Church of What's Happening Now.**

Control charts are in many ways like Geraldine because the most important data point for a control chart is the most recent event or, as Geraldine would say, "What's Happening Now!" Conditions and changes in the past cannot be resolved and fixed but we can act upon the current situation.

Control charts must be reported in a timely matter to allow us the opportunity to determine the reason why the change has occurred. Control charts demand that all results must be posted immediately.

Tactical Support for the Operator

In a manufacturing setting, the operator is our tactician. The operator must use the available means to its maximum and optimum in order to run the facility as correctly and consistently as possible. A separate set of tools which alert the tactician to change are required to support the tactician's mission of consistently running the available facility. These statistical tools, based on averages and the use of standard error, are called **confidence intervals** and will be the tactician's tool kit. The formula for confidence intervals is

Confidence Intervals

$$X_i = X_{Ave} \pm z * Std_{Ave}$$

OR

$$X_i = X_{Ave} \pm z * \frac{Std_{Ind}}{\sqrt{n}}$$

Confidence limits which detect change become our means of building control charts. The control chart uses a special kind of confidence limit called control limits. A **control limit** is an alarm that the product or process has changed from what has been routine for this machine.

What Makes A Control Chart? A control chart is the alarm system to note changes in a product or process function. The control chart monitors the various quantitative aspects of the function. This is a pictorial change alarm system.

The information to be plotted is comprised of two components: data and alarms.

Data or values to be monitored over time. Since we are looking only for an alarm of a shift or change, then the data will be massaged through averaging to assure a normal distribution.

Alarms to detect change. The following are the alarms that are used in control charts and are based on the fact that the data will be a normal distribution.

Average 50% of the data points above and below the average.

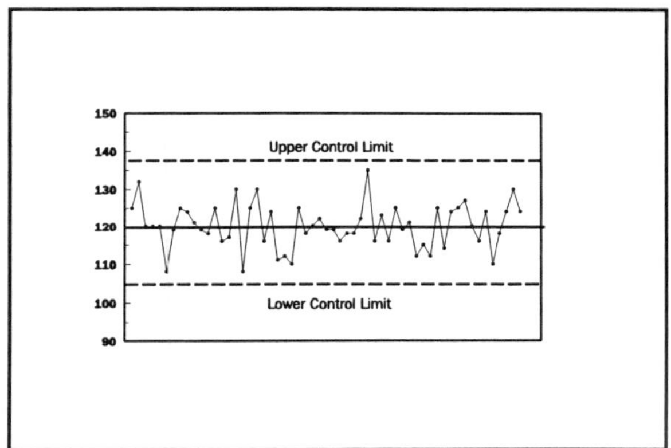

Zone 1 68% of the data points will fall inside the plus or minus one standard error interval.

Average + 1 Std_{Ave}
Average - 1 Std_{Ave}

Zone 2 95.4% of the data points will fall inside the plus or minus two standard error interval.

Average + 2 Std_{Ave}
Average - 2 Std_{Ave}

Zone 3 99.8% of the data points will fall inside the plus or minus three standard error interval.

Average + 3 Std_{Ave}
Average - 3 Std_{Ave}

More zones could be created but the possibility of having values in larger zones is so small that there is no need to include them.

A control limit is a special kind of confidence limit used for the specific purpose of assisting the operator run a machine consistently. A control limit was defined in the 1930s with a Z-value of three. This Z-value was selected such that any data point outside the control limits would be a rarity. The control limits are as follows:

$$Control\ Limits\ =\ X_{Ave} \pm 3 * Std_{Ave}$$

The original developers of SPC, defined the term control limits to mean a specific idea. Since control limits were defined so many years ago, any deviation from the definition is absolutely forbidden with no exceptions. This would be like trying to redefine the color red. We might be able to convince a few people to call red by the name of *blue*, but anyone to whom we had not talked would be extremely confused and rightly so.

In addition, **ALL** interpretation rules are based on control limits that are computed with a Z-value of three. Any changes would require complete revisions of these interpretation rules. No matter why, we have too much invested in our proper control limit definition to allow any changes or deviations. **Thus, control limits must be set only at a Z-value of THREE.**

Normally only the control limits appear on the control chart because more lines will create too much clutter for easy use of the control chart methodology.

We will discuss in Chapter Six the details of the calculations for the alarms associated with control charts. We will see some short cuts that were invented back in the 1930s to expedite the calculations of control limits and, most importantly, discover why these short cuts work.

What is the Management's Role? The concept of making what the customer wants is very important; we must be customer driven. At the same time, we can only evaluate capability when we are using the correct tool. Management must stay focused on the strategic needs.

Management must understand its responsibility for providing a facility that will make what the customer wants. Management must understand that it will be held accountable for this mission of the capability of the machine. Like we provide very targeted tools to support operations' mission, tools must be provided that assess management's mission of providing facilities that can meet the customer's needs.

Capable: When we say that we are capable, we are only talking about management and customer issues. The term *capable* means that the facility is making what the customer wants. There is a difference between the correct and consistent running of the facility and how well we are meeting our customer's expectations.

Management Decisions: When we understand the concept of capability, we realize that the objective of management is to assure that we are providing a facility that can make what the customer wants.

As management makes **strategic decisions** to plan for a goal, the limiting factor of availability that was placed on tactical decisions is removed. Management must find the *means* or facilities to supply operations. Strategic decisions involve determining the time, place, and resources required to accomplish our goal just as a general must obtain the resources, determine the place for the battle, and pick the time of the battle.

Strategic decisions are categorized as the defining and obtaining of the appropriate resources and then determining when and where to use them.

Management => Strategy
Management uses strategies and strategic decisions to continually evaluate the adequacy of the facility in making what the customer wants. This strategic decision is based on the tool we mentioned above, the capability study. If the facility is not capable, a strategic decision must be made concerning a proper course of action.

Which Tools Support the Management/Sales/Customers of a Continuous Operation?

There are many strategic analytical tools to support management/sales/customer decisions. One of these, **a capability study**, assesses how well what we are producing matches what the customer wants. A capability study would evaluate how well the product meets our customer's expectations. These studies are for a continuous process and assume the process is in control. This chart is an example of a Capability Chart. Use a graph like this for management reports.

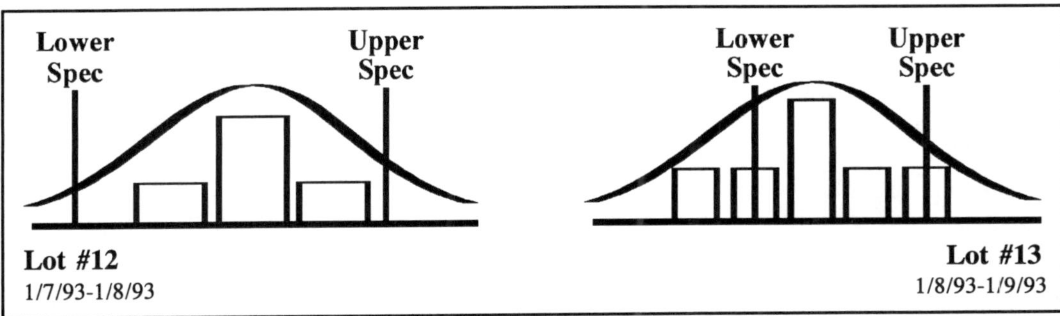

Because strategic decisions involve defining and obtaining appropriate resources and then determining when and where to use those resources, the determining of the facility is clearly a management/sales/customer function. The capability study gives management the information it needs to make appropriate decisions.

The archer assesses how the arrows are hitting the target and how many of them have hit the bull's-eye. As our chart shows, improvement is required to become a master. *Control* does not mean *capable* and *capable* does not mean *control*. *Control* and *capable* are two independent functions. When we manipulate them with one another, spectacular things will happen! Later we will see how they form a team.

Strategic Support for Management

In Chapter One we discussed the use of average, standard deviation, and the Z-value to build an interval. This basic idea of intervals will be used for two sets of analytical books to support Division of Labor.

Management decisions are strategic and assess impact. Management, or strategy, selects, assesses, and provides the means. All strategic analysis must support this mission. Since our management must assess the impact to the customer, customer specifications and the individual values must be used.

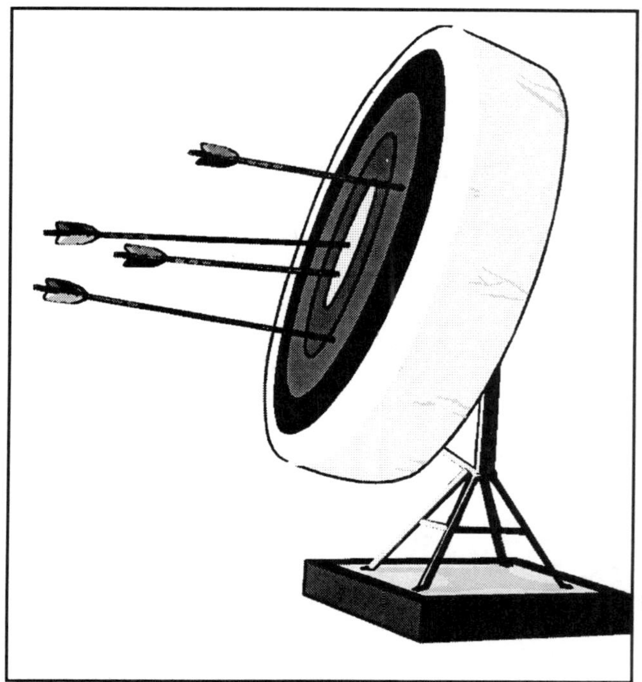

Establish Capability

In a manufacturing quality setting, strategic issues deal with assuring that we have the correct facility to make what the customer wants. Thus we must compare our data to the customer's specification. To accomplish the strategic mission, the individual values must be used. To make the objective a more general statement, the strategist is responsible for providing the correct facility and also assuring that any change to the facility is capable.

Prediction intervals are the statistical tools based on individuals and the use of standard deviation. All the statistical tools based on prediction intervals will be the strategist's tool kit. The formula for prediction intervals is

Prediction Intervals

$$X_i = X_{Ave} \pm z * Std_{Ind}$$

We will discuss in Chapter Eight the details of capability study calculations.

Statistical Process Control Team SPC is comprised of the team that includes both control charts (for running the facility) and capability studies (assessing the impact). When working together and in harmony, the team is an unbeatable combination.

The SPC Team

All changes are quickly detected by operations and then immediately reported to management. While operations is investigating to find the cause of the change, management is assessing the impact of the change on the customer. When both the cause and the impact are determined, then and only then can an action plan be plotted.

The chart to the right shows candy weights at the time of production. The weight of the candy has shifted from time period A to time period B. A second shift is shown from time period B to time period C.

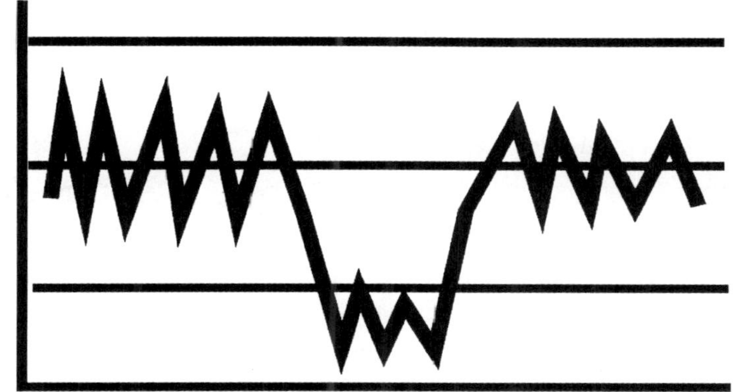

GooGoo Weight by Time of Production

The control chart below keeps us from repeating the time-series grouping mistake that so often is made in capability studies. Only data that is produced during a time of statistical control can be grouped together in a capability study.

When we analyze the candy weights being produced for time period A, the capability study on the next page results.

A	B	C
2 Day	2 Hr	2 Day
↔	↔	↔

The histogram to the right is the results from time frame A with the specification limits overlaid. This gives us a look at how the candy production for time frame A would meet the customer's specifications.

A = 0%

As we look at each time period, the following results appear.

A + B + C = 1%
A = 0%
B = 6%
C = 0%

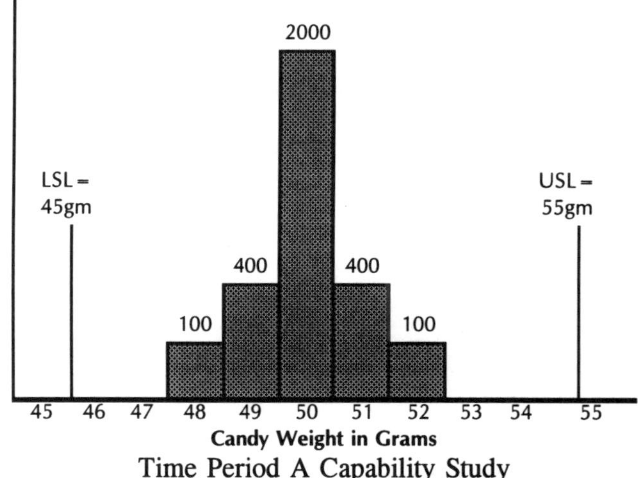

Time Period A Capability Study

The defective percentage for time period A is zero, for time period B is six, and for time period C is zero. Two distinctly different groups can be seen. If we attempted to group all the time periods together (A, B, and C), we would compute a defect rate of one percent which would be completely misleading and erroneous.

The tool to analyze this situation is called Statistical Process Control (SPC). SPC is the analysis tool of choice for a continuous operation.

The SPC team detects all changes and allows us to keep the improvements while getting rid of the deteriorations. All changes no matter how **small** are addressed with equal fervor. Thus a series of very, very minute shifts of deterioration do not slowly accumulate into one large problem.

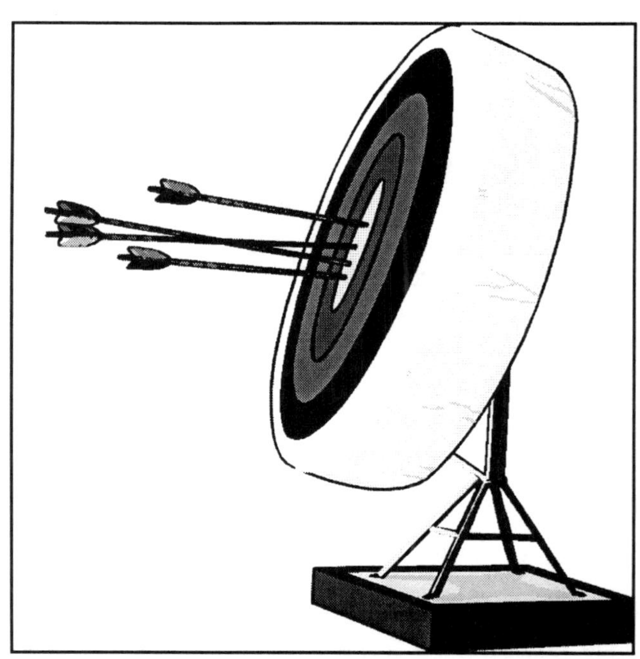

Continuing Improvement
...Through Control and Capability

Chapter Summary

Statistical Process Control SPC is the set of tools to monitor and control a continuous process. Control charts are the tactical tools to assist operations in the correct and consistent running of the facility. Capability studies are the strategic companion of control charts. They assess the impact of all changes on the customer so that management is never caught off guard.

The combined tools of control charts and capability studies allows our efforts to be focused to improve the product and the process. The SPC team can now support both strategic and tactical decisions with very specialized decision-making tools.

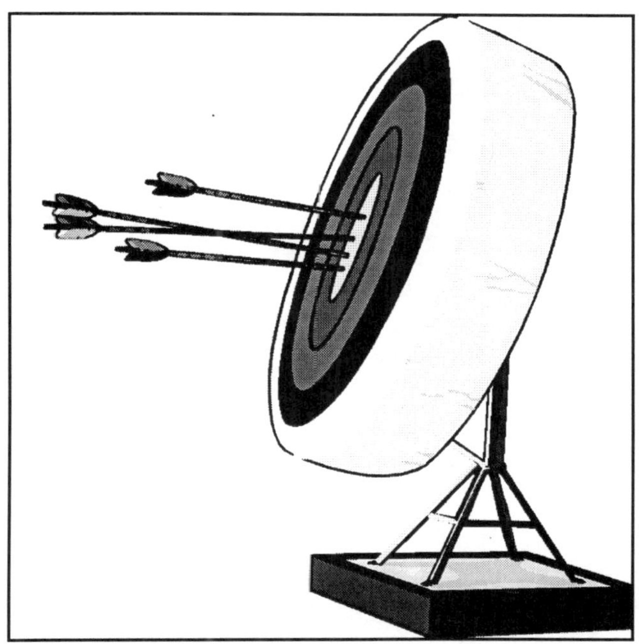

Continuing Improvement
...Through Control and Capability

The SPC Team

Control Charts The control chart is the operator's friend in accomplishing the tactical role of correctly and consistently running the machine. The following summarizes this chapter's explanation of a control chart and provides guidance as to which situations require this tool.

Consistency Alarm System Checklist

- Function of the Control Chart
    - Operator's Tool
    - One Particular Machine
    - Continuous Process
    - Time Production Oriented
    - Zones - (based on Bell Shaped Curve)
        - ✓ Average
        - ✓ Average +/-1 STD
        - ✓ Average +/- 2 STD
        - ✓ Average +/- 3 STD ! !
        - ✓ UCL - **Upper Control Limit**
        - ✓ LCL - **Lower Control Limit**
    - Recalibration
    - COWHN

Remember that control chart methodology is a charter member of the Church of What's Happening Now, commonly referred to as COWHN!

The chart to the right is an example of a control chart that has detected a shift for operations to deal with. Remember, a **customer's requirements or specifications have no place on a control chart**.

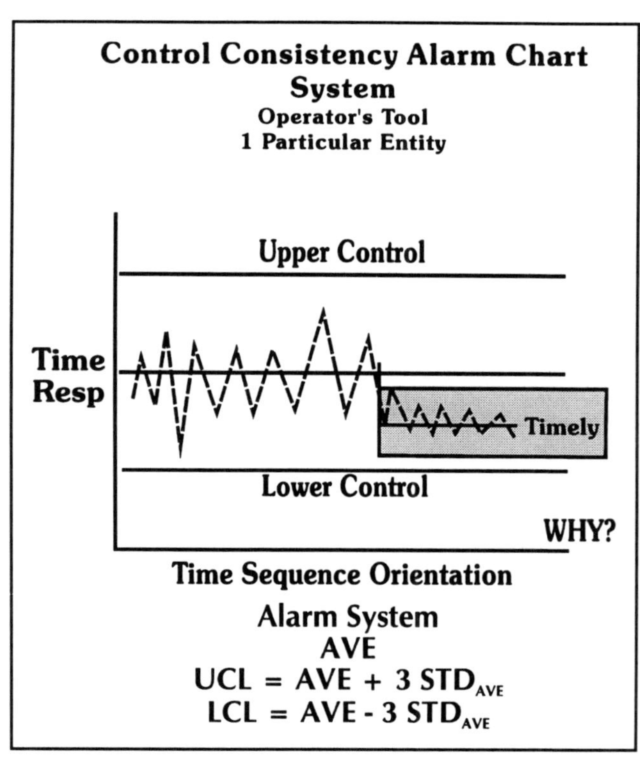

Capability Studies
The capability study is management's tool to assess how well the facility meets the customer's requirements. The capability study must work with the control chart in assessing all changes to determine for management the impact of the change on the customer.

SPC Team Control charts and capability studies form the dynamic team of SPC that is on a continuous search for improvement which may be retained and deterioration which can be eliminated.

The chart to the right is an example of a capability study which is used to assess the impact on the customer. Customer requirements and specifications belong on this tool.

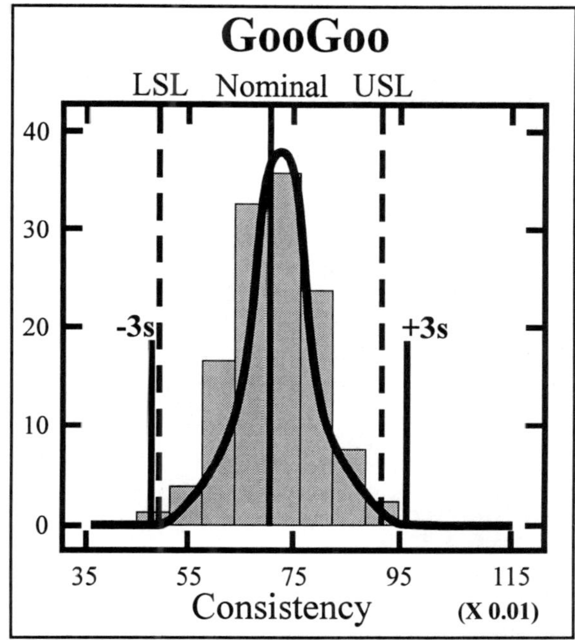

This chapter provides a general look at control charts and capability studies. Chapter Six will explain in detail the math for computing them. Chapter Eight will explain the mathematic computations for capability studies.

Chapter Five: Where Do We Start?

Mission

To assure that control charts and capability charts are being used correctly, their purpose must be stated clearly. Furthermore, this statement of purpose should be placed directly on the control chart. All participants (management, line supervisors, and most importantly, operators) should understand the purpose of the charts and how these tools will provide information to improve the process.

If *all* parties understand why we have control charts and why we are using them on a particular process, all parties will be focused on the mission of **continuous improvement**.

If all parties are not using the tools correctly and monitoring the proper metrics, we will make inappropriate decisions. Not only will we not make improvements, but the reverse, deterioration, will occur. We must understand the first principle of process management in order to focus on the right target.

A metric blueprint is a prerequisite for improvement to occur. A metric is a study area for which the blueprint gives focus and direction. The metric blueprint answers the question, "Are we monitoring the right thing?" This will assure that not only are we working hard, but also that we are working on the most important areas. The tool that contains the metric blueprint is called the **Walkabout**™. It will be explained later in this chapter.

In my book ***Become the Manager of the Future***, I told the story of my mother learning the art of biscuit-making from my grandmother. Since Grandma was an accomplished biscuit chef but my mother was not, the knowledge transfer was very difficult. Without the proper analysis tools used in the correct areas, Mama burned many biscuits and the baking knowledge was never transferred. My story showed how improper focus keeps us from making improvements. Let's give a brief summary and recap of the First Principle of Process Management.

First Principle of Process Management: **A fundamental understanding of BOTH the product and process is essential to improvement. Both the product and the process must be described and understood individually and separately. The underlying component for improving the product is the process.**

The First Principle of Process Management is crucial to improvement. There are two broad areas: **product** and **process**. We must have data from each area, and in order to improve the product, we must understand the process better.

A product is...

- **Noun**
- **Result**

When we are discussing products, we are focusing on the *result* of our activities. The product is always a noun.

Product

A process is...

- **Verb**
- **Activity**
- **Something that builds the product**

Process discussions are focused on the *activities* that build the product. The process is always a verb.

Process

WALKABOUT® Dependency Diagram

Area

1 Product Definition

✓ Use clear succinct paragraphs
✓ Do not use "buzz words" or acronyms
✓ Use nouns

Biscuits:

Flaky, Light, Round Bread

2 Process Definition

A. Mixing: *The proper amounts of ingredients mixed and then cut into dough.*

B. Baking: *The dough is heated for a set duration.*

Since the product is the goal and objective, many people monitor only the product, but to achieve the goal, an equal knowledge of the process is required. A clear understanding of the first principle adds focus to what we monitor. The importance of product and process knowledge must be clear because metrics and quantitative measures are vital to understanding the product and process. We will learn how to identify and measure metrics from the tool called the Walkabout™.

The front side of the Walkabout™ to the left gives us the techniques and facilities for documenting the product and process. The Walkabout™ provides the structure so that we will never skip a step or forget a component. The first step is to define and document the product. The second step is to define and document each step in the process. We write this information on the Walkabout™ shown here.

With this information, we have started on our journey of improvement. Many people have difficulty making a clear distinction between *product* and *process*. The important issue is to monitor both with equal fervor. Over-emphasis on either one will result in mistakes.

Importance of Process Data. When we monitor the product only, we are aware only of its deterioration or improvement. However, this awareness of the change of the product does not give us a *reason* for the change. Thus product monitoring is important, but it is:

- Reactionary
- Tolerates waste with no understanding of cause
- Accepts improvement with no understanding of cause

Without a knowledge of "why," we cannot improve. For us to make Grandma's biscuits, we must understand why things occur the way they do. Product information gives no insight into why we made good biscuits or why we burnt the biscuits. Since we don't know why we are making good biscuits, we cannot repeat the activities that produced them. **This loss of knowledge is tragic.**

When we monitor the process *and* the product, the process parameters allow us to know the conditions that caused the change. The monitoring of both product and process is essential to:

- Proactive decisions
- Improvements
- Elimination of waste
- Understanding the causes of change

Statistical Process Control, commonly referred to as SPC, is the tool to monitor both product and process metrics. Note even the name of SPC is *process*. This should keep our focus on the process to improve the product. Too often we only use SPC to monitor the product as if the title is Statistical *Product* Control. Remember, use SPC (process) to monitor both the product and the process metrics.

The knowledge of "why" allows us to improve continuously. By adding knowledge about the activities required to make a biscuit (process or oven), we begin to learn what is required to make good biscuits. The additional information about the different processes that were used for good biscuits and defective biscuits allows us to continue to repeat the improvements and to discontinue the defects.

In the perfect world of my grandmother and her biscuits, I never saw her check the biscuits when she removed them from the oven. She kept her focus on the process. Since she knew all the subtleties of baking, she took corrective action as the biscuits were baking in the oven, so that, in a proactive manner, she avoided ever producing any burnt biscuits. She was a perfect chef; but unfortunately, most of us are *not* perfect chefs of *our* processes.

Often we spend a great deal of effort monitoring defects. However, when we are talking about defects, we are actually discussing the product. The problem with focusing on these undesirable product characteristics is that once the information arrives, it is history because the product has already been produced. If a defect is present, we now have waste, unhappy customers, scrap, and re-work.

Since we live in a not-so-perfect world, and our knowledge of our process is not nearly so complete as my grandmother's was about her oven, we *must* continue to monitor both our biscuits and our oven with equal intensity. More precisely, we *must* monitor both process and product.

As we gather data and knowledge, we must accumulate equal amounts of knowledge about *both* the product and the process. This will allow us to build that knowledge base of both the product and the process.

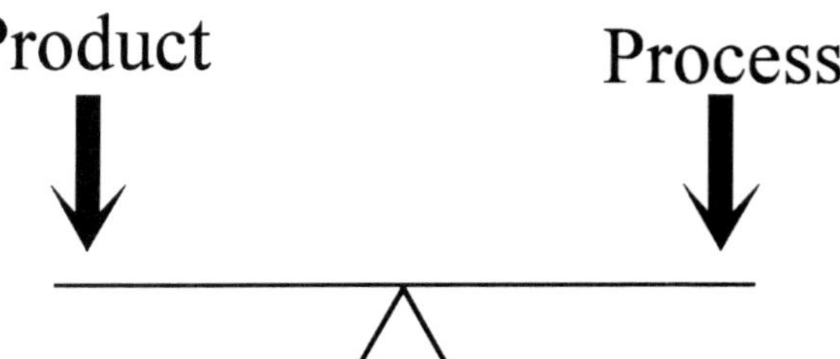

The first principle explains the interconnection between product and process. We cannot improve the end product without improving the process, and in order to improve the process, our knowledge base must be dramatically increased. In this chapter, we will draw a clear picture of the product/process relationship as a system of interacting and intertwined components. To simplify our efforts, we will use simple techniques to draw a schematic of the system. The system picture is called a **dependency diagram** and is part of the Walkabout™ we discussed earlier.

What Does the Walkabout™ Dependency Diagram Look Like? The chart below is the completed Walkabout™ for the baking of Grandma's biscuits. Let's use this example to explain how to read the Walkabout™. We discussed the first two sections of the front of the Walkabout™ earlier.

The dependency diagram is shown with circles for the products (think of southern biscuits) and rectangles for the processes (imagine an oven for baking). The schematic diagram shows the sequence in which each activity occurs.

Hanging from the products and processes are legs that have all the metrics (study areas) that we will want to study. The dependency diagram and the legs of the metrics form the metric blueprint.

The Walkabout™ provides a communication medium for each and every associate to learn, share findings from top to bottom, and retain this knowledge for current and future associates. This communication tool will allow us to do the following:

- **Learn** our product and process.
- **Share** our knowledge and what we have learned with our associates.
- **Teach** and train others in what we have learned.
- **Build a metric** and all its components through the use of the Walkabout™ as a metric blueprint.
- Gives us a **checklist** to assure we don't forget any of the details of our knowledge.

The Walkabout™ gives us *one common language and framework* to share and increase the knowledge that we will learn. Failing to have a common language leaves us in a chaotic state, but a common language gives us a method to get the best knowledge from all associates. Progress and knowledge *will not occur* if we don't have a common language.

The Walkabout™ describes our dependencies and defines metrics. With the Walkabout™ in place, we have a tool that will give us both a short and long range plan for metrics which will guide us in all our studies. This chapter will explain the following areas:

Metric Selection - Measurements and Targets to Support the Metric: Quantitative analysis is required for a vivid picture of the product and process. The term *metric* and the techniques required for defining metrics, measurements, and targets are explained in this section.

Using the Walkabout™ as a Metric Blueprint: The Walkabout™ is the comprehensive tool to evaluate how well the effort has been described and also becomes the tool to monitor compliance.

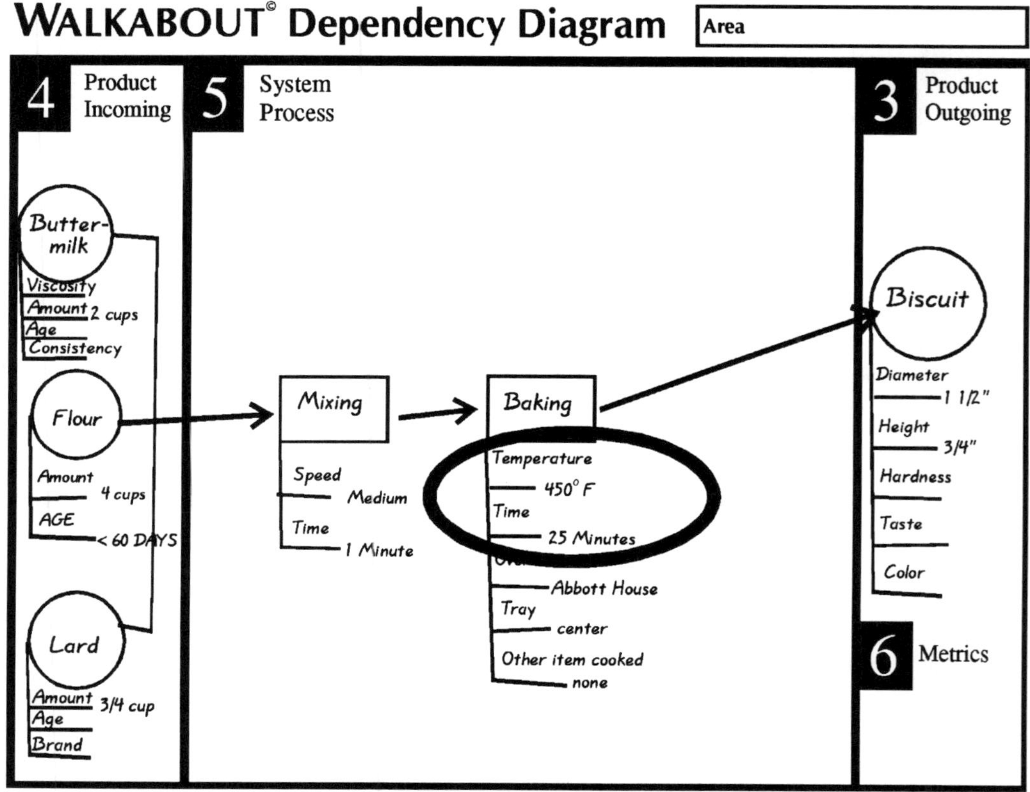

The Walkabout™ Dependency Diagram is a tool which allows us to make use of the First Principle of Process Management in order to document and improve our particular product and process. Again, the Walkabout™ Dependency Diagram documents the product and process and is driven by the First Principle of Process Management.

The Walkabout™ Dependency Diagram provides the information gathering required to document the system. The Walkabout™ Dependency Diagram is the tool to examine the system of incoming products, manufacturing processes, and outgoing products. It shows the sequence of processes and their dependencies. Metrics (study areas) are shown in relation to the products and processes.

Being able to see all the metrics in picture form provides a mechanism for identifying which metric to monitor. The Walkabout™, as a metric blueprint, provides the picture for selecting which metrics to start monitoring, which future metrics to monitor, and which measurements to determine. This allows us to understand and monitor our progress of current and future information needs.

My Grandmother used the Walkabout™ to clearly document the products, the process, and the product/process relationship. The Walkabout™ also contained all present and future product and process metrics, measurements, and target values and was first and foremost a tool to document how Grandma made the biscuits. Grandma used the completed Walkabout™ to teach my mother how to make southern biscuits.

When a metric is selected, we now have a place to begin our efforts and are ready to begin Statistical Process Control.

As Mama and Grandma worked to improve our biscuits, the Walkabout™ grew and grew. Both the process and product understanding are shown in the Walkabout™. This tool provides the medium to share this knowledge with future generations so that the legacy of southern biscuits will live on. We can see the Walkabout™ with the improved understanding that Mama added to our knowledge base.

We must strive to accomplish the levels of expertise in our process that my grandmother and mother achieved in their biscuit-making production. Grandma and Mama did not achieve their expertise by easy and lazy efforts. Improving our knowledge base is a worthwhile objective, but it ***requires hard work on the appropriate things.*** It is not enough to know what to do, but we must actually do it. We must build and use the tools of process management, use our knowledge and execute our plans.

Which Metric Do We Monitor First? Three issues are the driving forces for picking metrics for analysis. These issues are 1)the cost of monitoring the metric, 2)what and how much information is to be gleaned, and 3)the feasibility of measuring the metric.

These three issues will be the guide for us as to which metrics are monitored first. All our metrics will ultimately be monitored, but only so many can be monitored at once. Thus we must make strategic decisions as to which metrics to monitor first, second, third, and so on until all metrics have been monitored to give us complete knowledge of the product and process.

Good

As we pick our metrics, a balance of incoming product, manufacturing process, and outgoing product metrics must be maintained. These three must be kept balanced. A correct balance of information would result in a Walkabout™ as shown to the right, balanced or close to the same number of metrics for all areas. This will provide a tremendous amount of information for all areas. We can learn in a balanced way about both the product and the process.

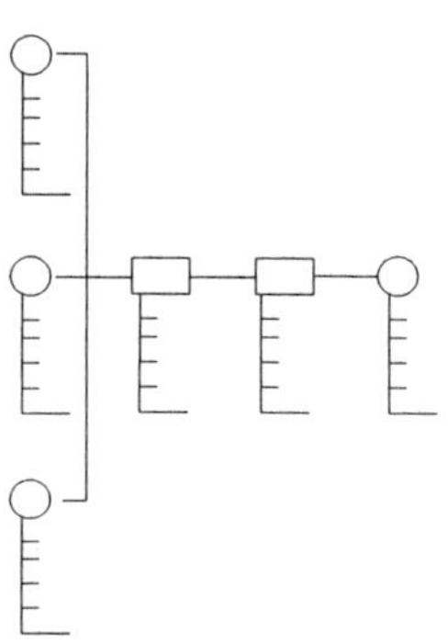

Remember to keep a balance of product and process metrics. This will give us accurate information that will facilitate our process and product improvement.

What did Mama choose as metrics for SPC? She knew that SPC was a tool to assist in the monitoring of a continuous operation. Let's see how Mama approached the selection of metrics and then used SPC for monitoring them.

Monitoring mama's biscuit making. Continuous operations are found in many situations. According to the completed Walkabout™, some of Grandma's metrics are continuous in nature. We will need to monitor these metrics with SPC, our tool of choice for continuous operations. Our completed Walkabout™ guides us by keeping the total context of all metrics so that we can pick our first monitor areas.

A **continuous operation** is one in which the same activities happen over and over again. Many operations fit this criteria. Statistical process control (SPC) is a set of tools that was designed specifically for this kind of situation. Now we must determine which metrics are continuous and thus a candidate for SPC. Our statistical tool selection process must move from *situation* to *tool*.

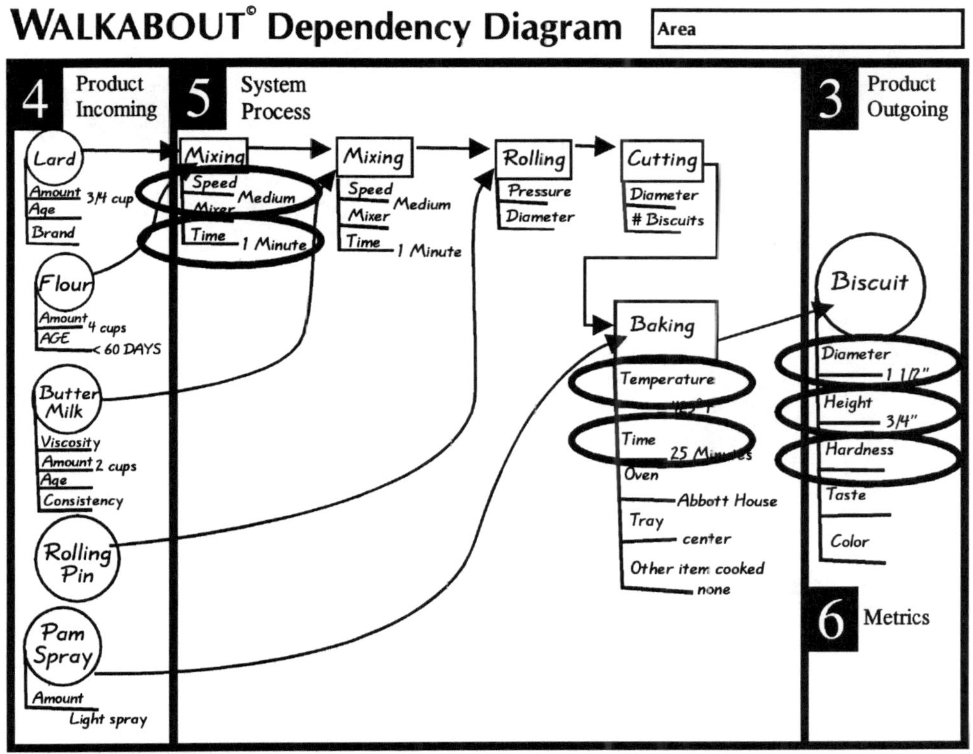

Oven Temperature. Grandma and Mama believed that oven metrics are important to good southern biscuits. Since the temperature may continuously change, a continuous surveillance method must be used to monitor it. SPC is the tool for the task of continuously monitoring the oven temperature.

Other biscuit and oven metrics may be candidates for SPC monitoring. These metrics include the mixer speed and length of mixing time; the oven duration; and the biscuit diameter, height, and hardness. These metrics would have control charts to assess the consistency of each metric. Each metric then could be compared against the specifications and targets to assess the capability of the process.

Once the metric is selected, we must learn to measure the metric.

How Will We Measure the Metric?

Measuring the metric is essential to the learning process. If we never shoot the bow, we will never know for certain how good our archery skills are. Similarly, taking readings is essential to our learning efforts. No measurements lead to no knowledge.

Measuring our metric is key to learning. Lord Kelvin said in 1883, "When you can measure what you are speaking about, and express it in numbers, you know something about it. But when you cannot measure it, when you cannot express it in numbers, your knowledge is of a meager and unsatisfactory kind."

When an archer shoots an arrow into a target, the different locations of the arrows in the target paint a picture about how he should improve. When the target includes numbered circles, the knowledge of the archer's accuracy is increased. Adding numbers to the target advances our ability to learn.

Walkabout™ as a measurement blueprint: The completed Walkabout™ will provide the blueprint for all measurement definitions.

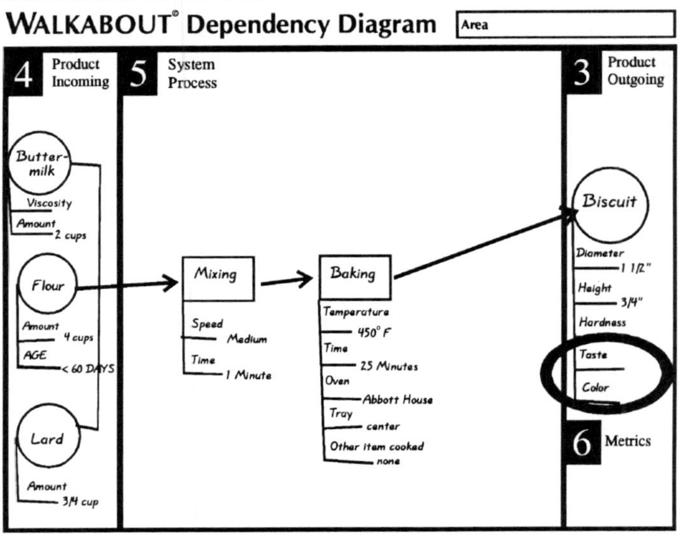

Some metrics will be either difficult or impossible to measure while other measurements may also by very expensive. In these situations, we do not want to lose our metric, but we may simply defer our measurement until taking it is feasible. With a blueprint of our metrics and measurements, we will never lose sight of our need to gather more information, and we will also know what information is missing. In the example above, technology is not available to measure taste or color. These are still metrics, but they will be added when the technology is available in the future.

What Measurement Types Are Available? There are two metric measurement types. The first type, called **attribute**, has Pass/Fail, True/False, or Counts as its properties. Attribute data is also referred to as **categorical data**. The second type has measurable properties and is called a **variable**. Variable data is also referred to as **continuous data**.

Attribute Data: Attribute or categorical data is Pass/Fail and Counts. This type of data is generally thought of in the passing of a part in manufacturing or in service industry (administrative) tasks. Even though this type of data is pass/fail, we can count the frequency of occurrence. Attribute data is "yes" or "no", "good" or "bad", "go" or "no go". Another example of attribute data is the count of the number of red M&Ms in an M&M bag.

How we measure the metric "flour age" can take on several forms. Attribute data could be one possible choice for determining the flour age by monitoring the number of bags produced before a certain date. The following describe how that could be accomplished.

All or Nothing: This most elementary type of attribute data determines whether an item is good or bad. We might count the number of biscuit batches in which the flour age exceeded a pre-specified number of days. If we set the target age to be less than 30 days, our measurement would then be the number of batches over 30 days old.

More or Less: In this next level of refinement, we compare our data to some set of benchmarks. An example is a comparison to a set of paint chips or a range. Grandma could use the more or less concept to identify when the biscuit dough is mixed to the proper consistency.

Sequence: In this example, we compare to a scale, such as 1 to 10. An example of this is a hardness scale in which a diamond is 10. Counting the number of batches within ranges could provide more insight into the age of flour. We could tabulate the flour bags' batches into ranges of age days.

Age Days
0 - 10
11 - 20
21 - 30
31 - 40
41 - ↑

Variable Data: Variable data is a continuous measurement. This type of data is thought of in terms that can be measured like temperature, length, width, diameter, shading, and days. Continuous data can be recorded on a scale: meters, grams, inches, or viscosity. The following describe how variable data can be expressed.

Units: Examples of this are length, width, time and weight. We might want to know the flour age in number of days since it was milled.

Multi-Dimension: We are now measuring against multiple units. An example of this is density, which is weight per volume. We want to know the flour age but rather than simply monitoring number of days, we want to divide the number of days by the number of biscuits in the batch.

Steps to Metric Measurements: There will always be several options available to us in our metric measurements. Consideration must be given to the amount of information to be gleaned, the cost of the measurement, the accuracy of the measurement, and the difficulty of the test procedure to accomplish the measurement. Over time, I would expect the technique for gathering the measurement to change and improve. These improvements should not be confused with more costs or just more precision; instead the improvements must increase product and process knowledge.

Determining the kind of measurement (attribute or variable) and the technique for gathering the measurements are key to increasing our knowledge base.

In addition to the obvious things like temperature and duration, metric measurements must be established for our cleanliness in housekeeping. For example, a frequency check for the cleanliness of the oven must be established because this is a vital metric measurement. Similarly, checking for incorrect operations must be a regular part of the housekeeping effort. It is reasonable then that these checks for incorrect operation must be included in the Walkabout™.

Activities for Metric Measurements

1. Determine the measurement technique and procedure for each metric. The intended use of our information will drive our test method. The need for different measurement methods to support operations versus protection of the customer should come as no surprise.

2. List the metric measurement that will be used for each incoming product, outgoing product, and process. Consider the following in determining the measurement.

   a. Amount of information to be gleaned
   b. Resource availability as a function of how much time the people have to do the measurement
   c. Skill of the people that will be asked to make the reading
   d. Cost of the measurement

   Once we have determined the measurement and the procedure for taking the measurement, we are ready to move to the next step and define the mission of the control chart.

3. What is the mission of the control chart? Remember that clearly defining the mission of the chart gives all associates an understanding of where the SPC effort is headed.

   ____

   ____

Having a clear mission for the control chart gives the tool purpose. With this purpose in hand, the chart has a start and an end because when we have learned all that we need about the mission, the chart may be no longer necessary.

Remember, the measurement process benefits from SPC techniques. Measurement capability is called repeatability, reproducibility and accuracy. Measurement Control is called *stability*.

How will we gather our measurements? Many people want to analyze all our data, but when *all* the products are measured, the cost of our testing will increase. A plan must be developed to monitor *enough* data but not too much. The testing plan must be both cost effective and accurate.

Population Statistics refers to the monitoring of all the data available for a set of information. For some situations, gathering all the data can be impossible or expensive.

Population statistics mistakes often are made when all defects are analyzed and this analysis is called a population simply because we are looking at all the defects. The information from this type of study is very biased and paints a very distorted picture. The defects are really a biased sample of the total population. What then is a sample?

Sample Statistics refers to the monitoring of a subset of information about a population while attempting to estimate the population statistics. In order to achieve our objective of estimating the population through a subset, bias must be avoided at all cost. Avoiding bias is done through a technique called **random samples**. Since random samples can be accomplished in many ways, the technique we choose depends on the situation.

For a continuous process, our objective is to detect any change as the process works. Therefore, chronological sequence is important to avoiding a biased picture. We must track all data in its time of occurrence.

Population = Sample + Error

The only way to avoid a risk of no sample error is to test all our product. But again, to test all the product is prohibitive. With this in mind, a certain amount of risk of getting bad information is possible. Later we will discuss the accuracy of our results.

When are readings taken? Since control charts are tactical tools and invoke central limit theorem to normalize our data, then clusters of readings will be captured. SPC is a tool for a continuous process and, with this in mind, our readings must be captured in chronological sequence. The readings will be pulled to form a grouping called a **subgroup.** The subgroups must maintain their time sequence and have equal amounts of time passed between each of the subgroups.

How are the individual readings of the subgroup pulled? Samples of product must not be drawn haphazardly and must not introduce bias toward either good or bad. As nearly as possible, a single subgroup must include product or process readings produced at the same time under the same conditions. In other words, the individuals will be consecutive readings if possible.

Remember our measurement readings are intended to paint a total picture of the process and product - not to just detect the bad product. This total picture must include central tendency, subgroup to subgroup variability, and reading to reading variability. In addition, it must detect change, monitor consistency, and assess impact to the customer.

This tool allows the continuous process to be understood and improved.

How often do I take a subgroup reading? Subgroup readings will be obtained as frequently as the process demands. As often as the operation changes, readings must be taken. This decision is dictated by the volatility of the operation and not by statistics. For instance, if we wish to detect variations in product due to worker "warmup" over the first couple of hours of a shift, we must measure product several times during those hours. Measuring once per shift will fail to detect the cause.

If the process distribution is stable (i.e. if the process is in control), we need only sample infrequently to detect changes in the distribution. If the distribution is not in control, we will need to sample more often to detect special causes.

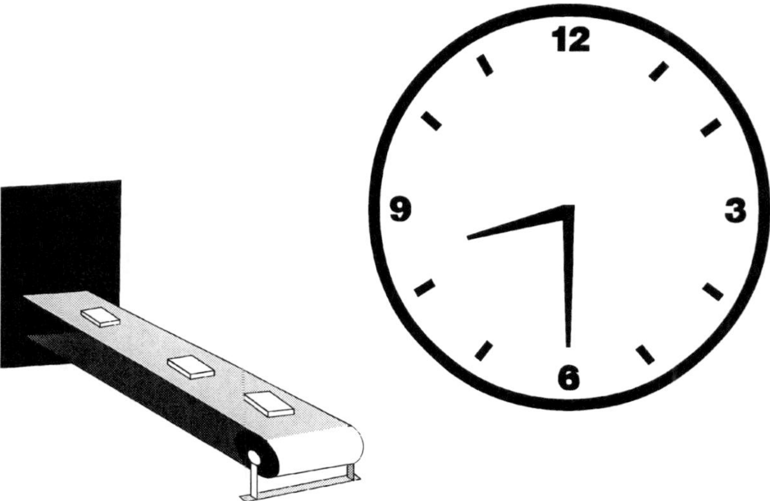

How many will be tested? Central limit theorem (the math trick we learned in Chapter Two) is the reason for our multiple readings in a subgroup. Central limit theorem is the tactical medium for making a non-normal distribution into a normal distribution. As we learned in Chapter Two, five samples in our subgroups will almost always guarantee a normal distribution. This assures simple alarm methods for the operator to use.

More samples in a subgroup will simply make the forced normal distribution a better normal. Generally, if more samples are to be pulled, we prefer a more frequent number of subgroups to more samples in a subgroup. As an example, instead of pulling 20 samples in a subgroup every hour, we would rather pull five samples in four subgroups every 15 minutes.

Remember the sample size is driven by central limit theorem, not by sampling theory or any attempt to reduce the risk in sampling error. ***As a rule of thumb, five samples are pulled in a subgroup if at all possible.***

Control Chart Misconception Many people think that control charts are a tool to assist the quality assurance (QA) department in weeding out all the defective product. However, the real use of a control chart is not for the quality assurance department but part of a total system of both control charts and capability studies to improve a continuous operation. Too often the control chart is used by the QA department, and because they have taken the initiative to use the tool, the operations management fails to get involved.

SPC cannot weed out defects but can only paint a picture of what is happening on a machine. This picture can be of tremendous value in the hands of operations to assess and build a plan of improvement. As such, the most asked question, "How many samples do I need to pull?" is not pertinent because once the process is in control, then all the values during this state can be collected in a time-series group for assessment.

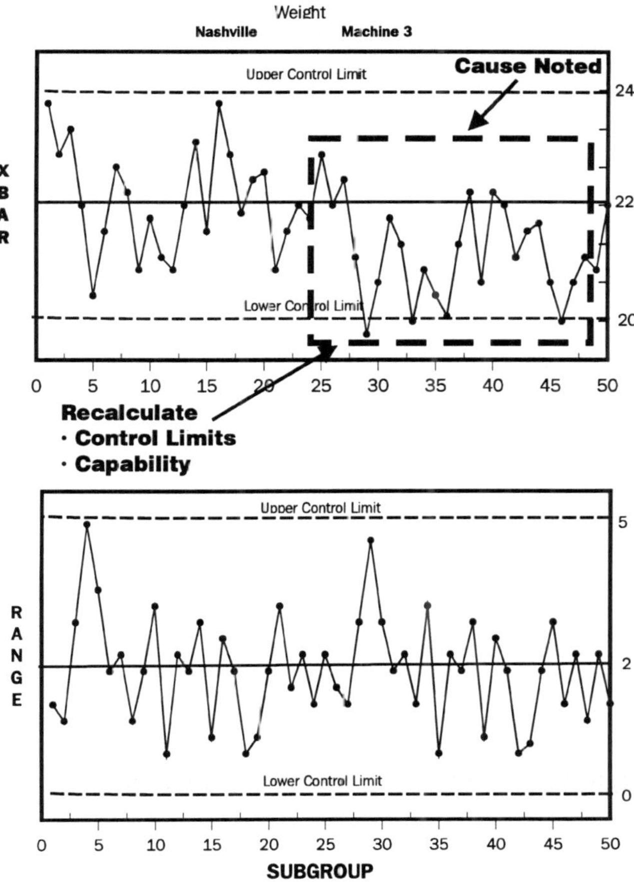

Accuracy Of The Capability Study An assumption of the capability study is that the process is in control (consistent). Since the capability study is based on a large number of subgroups that are all consistent, we would expect the accuracy of our data to give a very good picture of our product. In other words, we are looking at a very large sample of data to approximate the population. This large sample of data is comprised of all data during the time frame when the process is in control. The picture shows that we are analyzing our data over time.

The longer we are consistent, the more comfortable we are that our product decisions are based on good estimates. But remember, **SPC cannot weed out all the defective product**.

Control charts and capability studies must work as the SPC team to obtain the improvement results desired. Control charts detect change. Change can be either an improvement or a deterioration. Thus a control chart alarm must be assessed with the strategic tool of capability to determine the impact. The SPC team is powerful in its ability to learn the process and product of a continuous process.

Chapter Summary

This chapter gives us a direction for "Where to Start the SPC effort". First, a Walkabout™ is essential as a metric blueprint for providing guidance and balance to our metric monitoring selection process. This gives us tools to assure that we are monitoring the right and most important metrics. From our understanding of the first principle of process management, equal focus is placed on both process and product metrics.

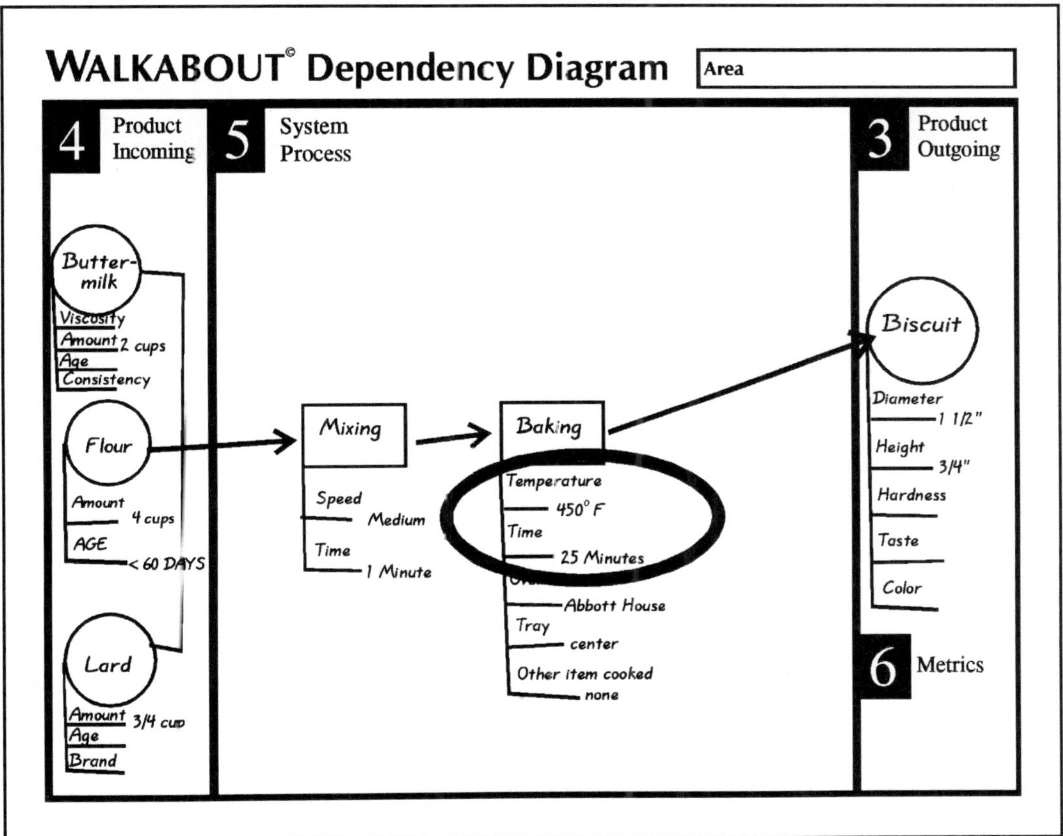

We determine which metric to monitor first based on the cost of monitoring, the information desired, and the feasibility of measuring.

Metric measurements may be expressed as attribute data (categorical) or variable data (continuous).

The steps for measuring each metric are established. Finally the how, when, where, and how many of the sample population is established for the SPC process.

With these issues under our belt we are ready to begin our SPC efforts. The next chapter takes us into the details of SPC.

Chapter Six: $\bar{X}$ And R Chart Construction

Statistical Process Control is comprised of 1) control charts and 2) capability studies.

A control chart is the operator's tool to monitor a particular machine or system over its time of function. It includes an alarm system with a mechanism for detecting any type of shift or change. This alarm must be dynamically re-calibrated to reflect the current status. The control chart supports the variable and attribute types of data. The variable control charts are $\bar{X}$ and R, $\bar{X}$ and s, and individuals. The attribute control charts are NP, P, C, and U.

A capability study is the management tool to assess how well the product we produce matches the product the customer desires.

To make our learning efforts easier, we will narrow our scope of interest to include only the operator and the tools that are designed to assist him. With this approach in mind, we will focus all our attention and work to covering **control chart methods** in detail. We will first cover the most used control chart: $\bar{X}$ and R charts. This chapter will cover the construction details of the $\bar{X}$ and R.

To support this chapter, Chapter Seven will cover the interpretation, reading, and use of the $\bar{X}$ and R. Since all control charts are based on a normal distribution, coverage of the $\bar{X}$ and R interpretation should make all other charts' use and reading very simple. All control charts use the same set of alarms to detect change.

Chapter Eight will cover the other two variable charts' construction. Chapter Nine covers the construction and use of attribute charts.

Understanding Capability by Implementing SPC

The overall objective of control charts' methods is to detect product or process change. The demonstrated effectiveness of a control chart is attributable to its objective and systematic analytical methods. *Shortcuts and "make-do" procedures lead to invalid conclusions and costly mistakes.*

The methods to effective control include the following steps:

- Calculate initial control limits
- Establish that the process is in control
- Continually survey the control charts for alarms (out-of-control conditions)
- Assess the capability of the in-control process

Calculate initial control limits To set the initial control limits, data must be gathered. As a guideline, thirty sets of values are captured to calculate and post on the charts. Remember, since a control chart is for one particular machine, the data must come from that one machine. Different control alarms for each machine would not be surprising. These same thirty subgroups are then used to compute trial control limits. The subgroup reading and the alarms are then posted on the chart for statistical control interpretation. This chapter will explain the mechanics and calculations for control limits.

Establish that the process is in control As each out of control condition is identified and dealt with, this procedure is repeated until no alarms have been detected. When the process is in control (no alarms have been detected), the control limits are locked in place. Chapter Seven will explain the reading and interpretation of the control charts.

Continually survey control charts for alarms (out-of-control conditions) Data is gathered to monitor against the alarm mechanism (control limits) for changes in the conditions of the machine. When an out- of-control situation or change is detected, the chart has done its job. The following tasks then must be undertaken:

1. Investigate the cause
2. Alert management to the change
3. Decide what to do

Chapter Seven will explain the details of the tasks associated with dealing with out of control conditions.

Assessing the capability of the in-control process The process that is in-control is then assessed for its suitability to produce product that meets the customer's needs. Chapter Ten will explain the calculations, assumptions, and use of capability studies.

Now let's focus our attention on the details dealing with variable data and the control charts that will monitor this type of data.

Variable Data Control Charts Variable data is continuous measurement of things like temperature, length, width, diameter, and time. A set of monitors exists for monitoring this kind of situation. From our coverage of basic statistical analysis, all three statistical components must be monitored: central tendency, variability, and distribution. Our control charts are the ongoing surveillance tools for a continuous process to alert us to any change in the statistical components.

Most control charts use the central limit theorem which eliminates the need to monitor the distribution. The exception is the individual chart which will be covered in Chapter Eight. A control chart displays changes over time. The charts then are analyzed in pairs to monitor central tendency and variability.

Variation: One chart displays changes in the variability of the measurements. The variability can both collapse or spread. The example here shows the variability collapsing. The solid line shows our original distribution and the dotted curve is tightened because of the variability change. This picture depicts the effect of the reduced variability.

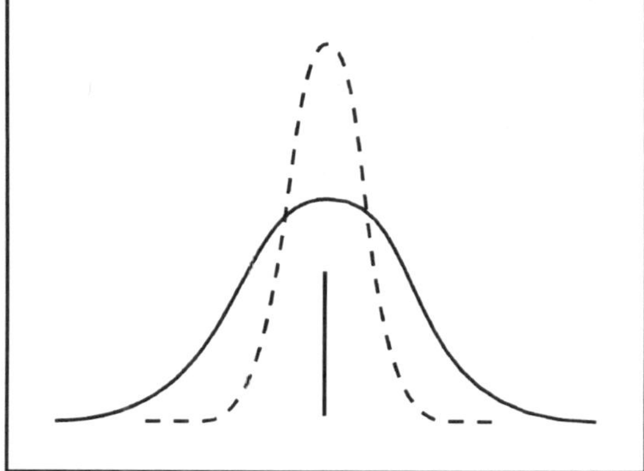

Variability Change

Average: A second chart displays the average values of the samples to detect shifts in the central tendency. The central tendency can shift up or down from its current position. In this example, the center has moved up from the current center position. The solid curve represents the original distribution and the dotted curve shows the center point of the distribution changing to a higher level.

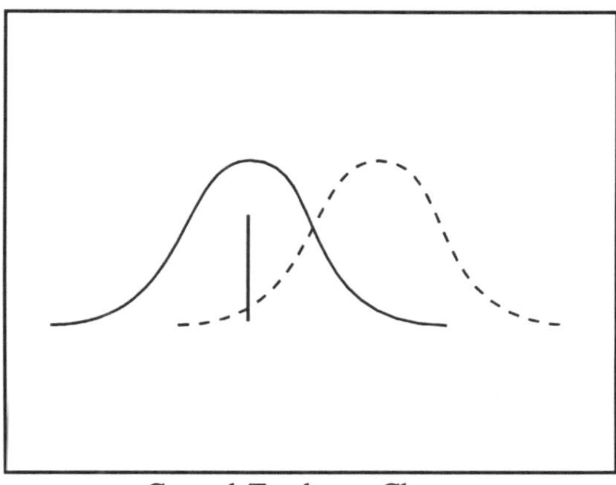

Central Tendency Change

Shifts in both the central tendency and variability can occur, and each can occur by itself. Control charts must be able to detect both the central tendency shift and the variability shift.

Now, we see in the graph that shifts in both central tendency and variability can occur simultaneously as well. If only one control chart attempted to detect both simultaneous shifts, the alarm could not draw attention to the particular area.

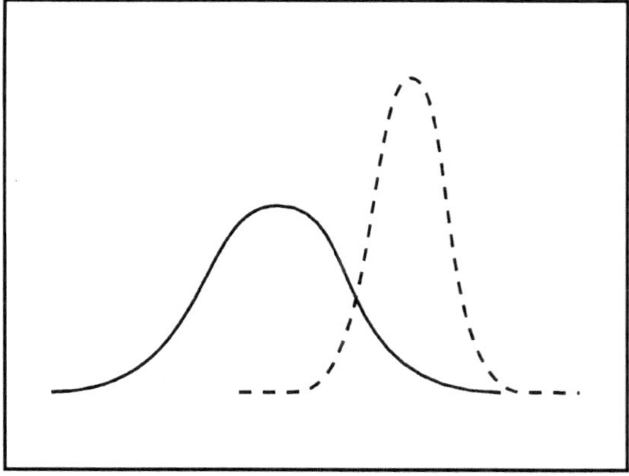

Both Central Tendency and Variability Changing

Let's return to our home smoke detector example in Chapter Four. The reason for having two smoke detectors, one in the kitchen and one in the bedroom, is to help reduce the search area for the fire causing the alarm. If the smoke detector in the kitchen is buzzing, then our first investigation is going to be in the kitchen. No one would start looking for the fire at the end of the house opposite from the smoke detector that is sounding. This reduces the area that we must search and investigate.

In like manner, we always need two control charts, one to monitor variability and one to monitor central tendency. This, like the multiple smoke detectors, will help us reduce our search and investigation area.

Now let's look at the pairs of charts that are classically use to monitor variable data.

What are the Classic Variable Control Charts? In the 1930s a series of control charts were developed to monitor variable data. Each set was designed to monitor for a particular situation. The three classic charts are $\bar{X}$ & R, $\bar{X}$ & S, and individual charts.

$\bar{X}$ & R are the prime charts for monitoring a high volume continuous manufacturing facility. The $\bar{X}$ & S charts are appropriate for the situation in which a large number of samples are in a subgroup. The last classic control chart is the individual chart for times when subgrouping is impossible or undesirable.

All of the classic control charts are comprised of two control charts. One chart monitors central tendency for any changes. The other control chart monitors for changes in the variability of the process. The table below shows what each chart is designed to monitor.

	Central Tendency	Variability	Distribution
$\bar{X}$ & R	$\bar{X}$	R	Central Limit Theorem
$\bar{X}$ & S	$\bar{X}$	S	Central Limit Theorem
Individual	X	Moving Range	Histogram

Remember, do not confuse control charts and control limits with a management chart to assess customer specifications. Control limits help us determine if a process is producing consistent output: that is, whether its distribution is stable and the process is in control and not changing. Later, we will apply specification limits to the distribution of a controlled process to determine its capability. Then, we will see that a process can be in control, but consistently producing defects.

Now let's focus on the $\bar{X}$ & R. This will allow us to complete all the calculations of one set of charts then move into Chapter Seven to discuss use of the control chart. This approach will allow us to cover one chart completely. The remaining charts then are continuations of this same approach. Chapter Eight will then cover the remaining variable charts and all the attribute charts.

We covered in Chapter Five, "Where Do We Start?", the details of how many samples, when to pull the samples, how to pull the samples, and how to measure the metric. Now we move on to the construction of $\bar{X}$ and R charts.

$\bar{x}$ And R Charts

This chart, the most commonly used, is geared to high volume manufacturing. Users typicically monitor product metrics such as the weight a biscuit, the diameter of a drinking glass, or the weight of a candy bar. This chart is used also for process metrics. This chart typically uses samples of five in a subgroup. Be sure that you plan for adequate resources to handle this large quantity of data.

When To Use: This set of charts is composed of the classic control charts. They are intended for high volume continuous manufacturing.

$\overline{X}$ Chart is designed to monitor the movement or changes in the central tendency of the process. Throughout this book the term $\overline{X}$ will be used for average of the subgroup and $\overline{\overline{X}}$ for average of the averages. The $\overline{X}$ Chart has the following plot lines:

$\overline{X}$ Chart Alarms

$\overline{\overline{X}}$	which is the benchmark of the central tendency of the process.
UCL_{AVE}	which is the upper control limit and is three (3) standard errors above the mean.
LCL_{AVE}	which is the lower control limit and is three (3) standard errors below the mean.

$\overline{X}$ Chart Data

$\overline{X}$	which is the average of each sub-group and becomes the data point to be monitored.

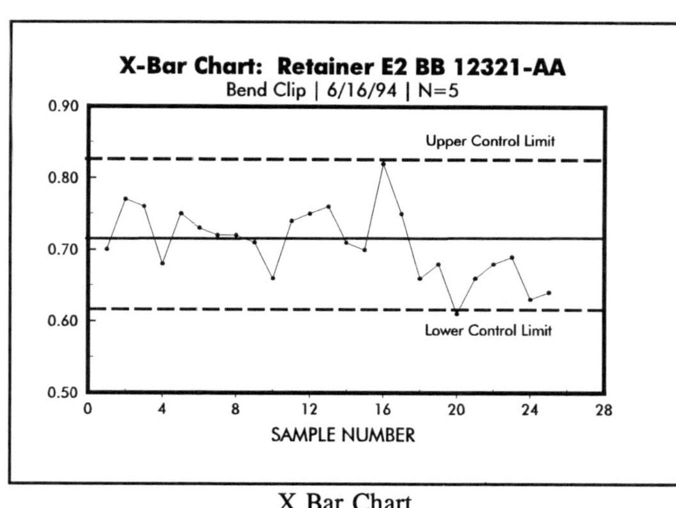

X Bar Chart

The above chart is an example of a $\overline{X}$ chart and the remainder of this chapter will explain how to compute the data and control limits for this chart.

R Chart is designed to monitor the movement or changes in the variability of the process. The R Chart has the following plot lines:

R Chart Alarms

$\overline{R}$	which is the benchmark of the central tendency of the process.
UCL_R	which is the upper control limit and is three (3) standard errors above the mean.
LCL_R	which is the lower control limit and is three (3) standard errors below the mean.

R Chart Data

R	which is the range of each sub-group and becomes the data point to be monitored.

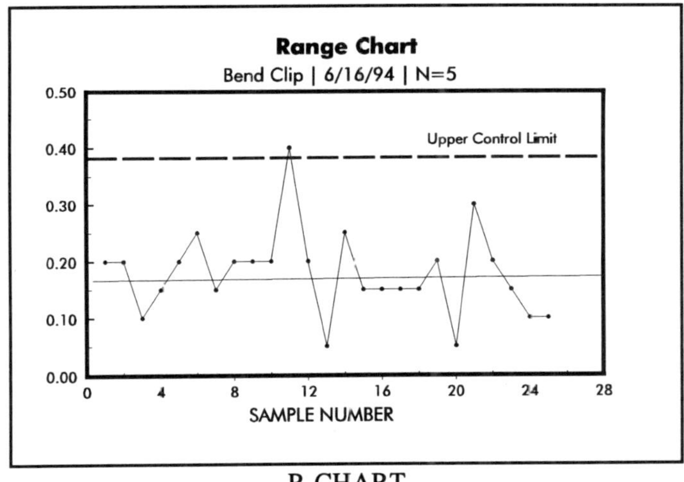

R CHART

The above chart is an example of a R chart and the remainder of this chapter will explain how to compute the data and control limits for this chart.

THE SPC Team The $\overline{X}$ chart and R chart form a powerful team with the $\overline{X}$ chart monitoring central tendency and the R chart monitoring variability. The prime objective is detection of minor shifts or change while the change is small and manageable. Now let's learn how to construct both charts.

For a simple text example, I have prepared the following product metric measurement data to analyze. Our manufacturing facility for this exercise is a high volume candy plant. Candy is coming from one machine down a conveyor belt in the sequence that it is produced. Each set of readings is taken consecutively and the subgroups are taken every hour. We have test data for the 1:00, 2:00, 3:00, and 4:00 time periods. This data will be used to generate our initial control limits.

This volume of data is only for this text and would not be sufficient volume to build a good set of limits. In the real world and in actual practice, larger volumes are required to provide a vivid picture of the process. As a rule of thumb, thirty subgroups would be required to start to determine the control limits.

SUBGROUP	1:00	2:00	3:00	4:00	5:00	6:00
SAMPLE						
1	10	14	10	10		
2	11	8	10	11		
3	13	10	9	9		
4	9	10	10	10		
5	10	9	10	10		

Looking in depth at the 1:00 readings of 10, 11, 13, 9, and 10. These are taken at precisely 1:00 and each reading is from the next five consecutive candy clusters. The actual readings are fed into the control chart mechanism. The data for control charts is the normalized values or averages. The $\overline{X}$ value is the average of the five readings. The R value is the difference between the high of the five readings (13) in the 1:00 subgroup and the low of the five readings (9) in the 1:00 subgroup. The 1:00 control chart readings of $\overline{X}$ and R are computed as follows.

$$\overline{X}_{1:00} = \frac{10+11+13+9+10}{5} = \frac{53}{5} = 10.6$$

$$R = 13-9 = 4$$

Now each one of the hourly subgroups must be calculated in the same manner. The results of these calculations is shown in the table on the next page.

$\bar{X}$ And R Charts Data The following are the algebraic formulas that were used on the prior page to compute 1:00 subgroup control chart data points. They are used to derive the data to plot for the $\bar{X}$ and R charts. Average data is used to provide a means to alert operators when a shift in the process has occurred. This should eliminate the potential for over adjusting control or eliminating a test due to an overzealous alarm.

$\bar{X}$ & R control chart data:

X_n is the data reading for the particular subgroup where n equals the size of the subgroup.

$$\bar{X} = \frac{X_1 + X_2 + X_3 + \ldots + X_N}{N}$$

where N equals the number of samples in the sub-group

$$R = | X_{Highest} - X_{Lowest} |$$

SUBGROUP	1:00	2:00	3:00	4:00	5:00	6:00
SAMPLE						
1	10	14	10	10		
2	11	8	10	11		
3	13	10	9	9		
4	9	10	10	10		
5	10	9	10	10		
Total	53	51	49	50		
Samples	5	5	5	5		
$\bar{X} = Total / Samples$	10.6	10.2	9.8	10		
R	4	6	1	2		

On a control chart, the actual individual candy readings will not be used. From this point on in this chapter, no use of the individual readings will be made. Therefore in the tables that follow, all of the individual readings have been eliminated. When we return in Chapter Ten to discuss capability studies, the individual readings will become significant again.

How are $\bar{X}$ Chart, Control Limits Calculated? Revisiting our definition of control limits, we remember that control limits are operational tools to detect a change and are the average plus or minus three standard errors. The upper control limit is plus while the lower limit is minus.

In a manufacturing setting, the operator is our tactician. The operator must use the available means to its maximum and optimum. The operator must run the facility as correctly and consistently as possible. The tool required to support the tactician's mission of consistently running the available facility is the control limits. This tool will alert the operator (tactician) to change that is unusual for his machine. Control limits are based on confidence intervals for a particular machine. The control limit formula uses confidence intervals and is as follows. Remember, control limits were defined in the 1930s with a Z-value of three. This Z-value was selected such that any data point outside the control limits would be a rarity. The control limits are as follows:

$$Control\ Limits = X_{Ave} \pm 3 * Std_{Ave}$$

Bringing the relationship of standard deviation and standard error forward, we know that

$$\frac{STD_{Ind}}{\sqrt{n}} = STD_{Ave}$$

Now, if we combine the formula above with the relationship of standard deviation and standard error, the following formula results.

$$Control\ Limits = X_{Ave} \pm 3 * \frac{Std_{Ind}}{\sqrt{n}}$$

Since control limits were developed in the 1930s for operators, the inventors knew that asking an operator to do all the math in the above equation was an impractical task. Obviously, some simplifications were required. This necessity drove the development of a series of very good approximations that made the math requirements less complex. Standard deviation is a rather laborious calculation with all the ranges, squaring, and summing. An approximation for standard deviation was found using the range value. The formula is show below. Range is not only an easier number to compute but also an easier number to conceptually understand.

$$Factor * Standard\ Deviation = Range$$

Remember, all these are *approximations*, not exact values. As a matter of fact, all statistics are made of numbers that are an approximation of the exact number. Standard deviation is an approximation of the variability of the process, not the exact value. Therefore using approximations does not detrimentally affect our monitoring of the process.

The prior equation can be rearranged to give the equation below.

$$Standard\ Deviation = \frac{Range}{Factor}$$

Using our naming convention for standard deviation, the equation reads as follows

$$Std_{Ind} = \frac{Range}{Factor}$$

Returning to the original equation for control limits as shown below:

$$Control\ Limits = X_{Ave} \pm 3 * \frac{Std_{Ind}}{\sqrt{n}}$$

This equation can now have the standard deviation (Std_{Ind}) replaced with the range divided by a factor equivalent value. This replacement is shown below:

$$Control\ Limits = X_{Ave} \pm 3 * \frac{\frac{Range}{Factor}}{\sqrt{n}}$$

The interval piece of the equation is composed of the number "three", the square root of the number of samples in the subgroup, and the range standard deviation conversion factor. To reduce the operator's workload then, the mathematics can be done one time and inserted into a table that only changes as the size of the subgroup changes. The table is shown below:

Subgroup Size	Factor A_2
2	1.880
3	1.023
4	0.729
5	0.577
6	0.483
7	0.419
8	0.373

Notice the factor is called A_2. The A_2 gets smaller as the subgroup size gets larger. This happens because the factor A_2 is derived from a group of constants with only the square root of n changing. Since the square root is our dividing number, the A_2 will get smaller as this number gets bigger.

As the control limits are plotted on the control chart, no more lines are used because more lines will create too much clutter for easy use of the control chart methodology.

$\bar{X}$ Chart Formula

Data The following formula computes the data points to plot and monitor on the $\bar{X}$ Chart. X_n is the data reading for the particular sample in the subgroup where n equals the sample number.

$$\bar{X} = \frac{X_1 + X_2 + X_3 + \ldots + X_n}{n}$$

where n equals the number of samples in the sub-group.

Alarms The following summarize the alarms and the calculations to arrive at them.

Center Point

The next formulas compute the alarms or control limits for the $\bar{X}$ chart. The first alarm is the average of the averages as calculated in the following formula:

$$\bar{\bar{X}} = \frac{\bar{X}_1 + \bar{X}_2 + \bar{X}_3 + \ldots + \bar{X}_K}{K}$$

where K equals the number of sub-groups.

Control Limits The general equation for the control limits for the $\bar{X}$ chart has already been explained in detail. Now let's summarize all the elements on one page.

$$Control\ Limits = X_{Ave} \pm 3 * Std_{Ave}$$

or, using the constant derived earlier, the equation is now expressed as follows:

$$Control\ Limits = \bar{\bar{X}} \pm A_2 * \bar{R}$$

The R values for each subgroup were calculated earlier and these values are now used to compute the control limits. The next formula is used to compute the average R called $\bar{R}$.

$$\bar{R} = \frac{R_1 + R_2 + R_3 + \ldots + R_K}{K}$$

where K equals the number of sub-groups

This $\bar{R}$ is used to approximate and compute the interval around the center. The factor that was derived in the previous section will be used to compute the interval of three standard errors above and below the center.

The upper control limit is computed by adding three standard errors to the average. This interval is approximated by multiplying the R_{Ave} by the factor called A_2. The table below is used to find the appropriate A_2 value. First the number of samples in the subgroup must be known. This number is found in the first column labeled ***Subgroup Size*** and the A_2 value is then obtained from the first column to the right. This same methodology can be used to compute the lower control limits.

A_2 is the conversion factor to arrive at three standard errors for use in determining UCL & LCL of $\bar{X}$ chart.

The other factors on this table will be covered later in this chapter and in Chapter Ten.

Subgroup Size	A_2	d_2	D_3	D_4
2	1.880	1.128		3.267
3	1.023	1.693		2.574
4	0.729	2.059		2.282
5	0.577	2.326		2.114
6	0.483	2.534		2.004
7	0.419	2.704	0.076	1.924
8	0.373	2.847	0.136	1.864
9	0.337	2.970	0.184	1.816
10	0.308	3.078	0.223	1.777
11	0.285	3.173	0.256	1.744
12	0.266	3.258	0.283	1.717
13	0.249	3.336	0.307	1.693
14	0.235	3.407	0.328	1.672
15	0.223	3.472	0.347	1.653

XBAR and R CHART Conversion

For a subgroup sample size example of five, the A_2 value is .577. This A_2 of .577 is the most commonly used because of the high use of samples of five in the subgroup to normalize the data. This value is then inserted in the equations below.

$$UCL_{AVE} = \bar{\bar{X}} + A_2 \ \bar{R}$$

$$LCL_{AVE} = \bar{\bar{X}} - A_2 \ \bar{R}$$

where A_2 $\bar{R}$ is an approximation of three standard errors.

Now let's move back to our simple example and compute all the alarms and plot the chart.

Example alarm calculation

Center Point The following computes the alarms or control limits for the $\bar{X}$ chart for our simple example. First, the alarm of the average of the averages is calculated.

$$\bar{\bar{X}} = \frac{\bar{X}_1 + \bar{X}_2 + \bar{X}_3 + \ldots + \bar{X}_K}{K}$$

$$= \frac{10.6+10.2+9.8+10.0}{4}$$

$$= \frac{40.6}{4}$$

$$= 10.15$$

The R values for each subgroup are calculated and these values are used to compute the control limits. The next formula is used to compute the average R called $\bar{R}$.

$$\bar{R} = \frac{R_1 + R_2 + R_3 + \ldots + R_K}{K}$$

$$= \frac{4+6+1+2}{4}$$

$$= \frac{13}{4}$$

$$= 3.25$$

This R_{Ave} is used to approximate and compute the interval around the center. The factor that was derived in the previous section will be used to compute the interval of three standard errors above and below the center. The subgroup size for the example is five, the A_2 value is .577. This value is then inserted in the equations below.

$$UCL_{AVE} = \bar{\bar{X}} + A_2\ \bar{R}$$

$$=10.15+.577(3.25)$$
$$=10.15+1.88$$
$$=12.03$$

$$LCL_{AVE} = \bar{\bar{X}} - A_2\ \bar{R}$$

$$=10.15-.577(3.25)$$
$$=10.15-1.88$$
$$=8.27$$

Subgroup Size	A_2	d_2	D_3	D_4
2	1.880	1.128		3.267
3	1.023	1.693		2.574
4	0.729	2.059		2.282
5	0.577	2.326		2.114
6	0.483	2.534		2.004
7	0.419	2.704	0.076	1.924
8	0.373	2.847	0.136	1.864
9	0.337	2.970	0.184	1.816
10	0.308	3.078	0.223	1.777
11	0.285	3.173	0.256	1.744
12	0.266	3.258	0.283	1.717
13	0.249	3.336	0.307	1.693
14	0.235	3.407	0.328	1.672
15	0.223	3.472	0.347	1.653

XBAR and R CHART Conversion

The table below shows all the details of the example. As an exercise, plot the values on the blank chart at the bottom of this page and compare your work to the completed chart later in the chapter.

$\bar{x}$ Chart Example

Subgroup	1:00	2:00	3:00	4:00	5:00	6:00
Samples	5	5	5	5		
$\bar{X}$ = *Total / Samples*	10.6	10.2	9.8	10		
R	4	6	1	2		

$$\bar{\bar{X}} = \frac{10.6 + 10.2 + 9.8 + 10}{4} = \frac{40.6}{4} = 10.15$$

$$\bar{R} = \frac{4 + 6 + 1 + 2}{4} = \frac{13}{4} = 3.25$$

$$UCL_{\bar{\bar{X}}} = \bar{\bar{X}} + A_2\bar{R} = 10.15 + .577(3.25) = 12.03$$

$$LCL_{\bar{\bar{X}}} = \bar{\bar{X}} - A_2\bar{R} = 10.15 - .577(3.25) = 8.27$$

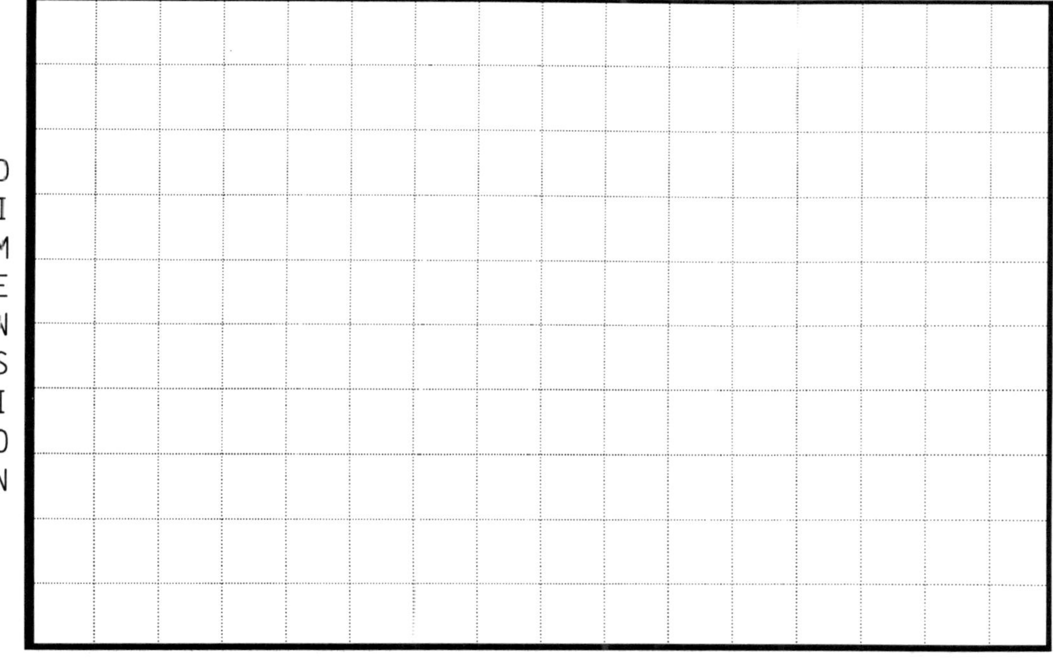

R Chart Formula

Data The following formula computes the data points to plot and monitor on the R Chart. R_n is the data reading for the particular sample in the subgroup where n equals the sample number.

Range = Maximum - Minimum

Alarms The following summarize the alarms and the calculations to arrive at them.

Center Point

The next formulas compute the alarms or control limits for the R chart. The first alarm is the average of the averages as calculated in the next formula. The R values for each subgroup were calculated earlier and these values are now used to compute the control limits. The next formula is used to compute the average R called $\bar{R}$.

$$\bar{R} = \frac{R_1 + R_2 + R_3 + \ldots + R_K}{K}$$

where K equals the number of sub-groups

Control Limits The general equation for the control limits for $\bar{\mathbf{X}}$ has already been explained in detail. Now let's take our knowledge from the $\bar{\mathbf{X}}$ chart and use it to compute control limits for the $\bar{R}$ chart. First the general equation for control limits is still the following.

$$Control\ Limits = X_{Ave} \pm 3 * Std_{Ave}$$

The $\bar{R}$ value is used to approximate and compute the interval around the center. The factor that was derived in the previous section will be used to compute the interval of three standard errors above and below the center of the $\bar{\mathbf{R}}$.

How can we compute $\bar{\mathbf{R}}$ upper control limits? Our first derivation will be of the upper control limit for the $\bar{\mathbf{R}}$ chart.

$$UCL = Ave + 3 * Std_{Ave}$$

$$= \bar{R} + Factor * \bar{R}$$

Now using algebra, the $\bar{R}$ can be pulled outside the parenthesis to allow the summing of the number "one" and a single remaining factor.

$$UCL = \bar{R}\ (1+Factor)$$
$$= \bar{R}\ (Factor_2)$$

The $Factor_2$ from above is called D_4. The upper control limit for the $_R$ Chart = (D_4) $(\bar{R})$ where (D_4) $(\bar{R})$ is an approximation of 3 standard errors. This leaves us with a much simpler equation for computing the upper control limit. Now the lower control limit must be computed separately because D_4 only works for the upper control limit. Finally the equation reads as follows.

$$UCL_R = \bar{R} * D_4$$

The table is shown below:

Subgroup Size	Factor D_4
2	3.267
3	2.574
4	2.282
5	2.114
6	2.004
7	1.924
8	1.864

Notice the factor for the upper control limit is called D_4. The D_4 gets smaller as the subgroup size gets larger. This happens because the factor D_4, like A_2, previously is derived from a group of constants with the square root of *n* changing. Since the square root is our dividing number, the D_4 will get smaller as this number gets bigger.

How can we compute $\bar{R}$ lower control limits? Our first derivation computed upper control limits. Now the lower control limit for the $\bar{R}$ chart must be calculated.

The $\bar{R}$ is used to approximate and compute the interval below the center. Another factor will be derived to compute the interval of three standard errors below the center.

$$LCL = Ave - 3Std_{Ave}$$

$$= \bar{R} - Factor * \bar{R}$$

$$= \bar{R}\,(1 - Factor)$$

$$= \bar{R}\,(Factor_3)$$

The $Factor_3$ is called D_3. The lower control limit for the $_R$ Chart= $(D_3)(\bar{R})$ where $(D_3)(\bar{R})$ is an approximation of three standard errors. Finally the equation reads as follows.

$$LCL_R = \bar{R} * D_3$$

The table is shown below:

Subgroup Size	Factor D_3
2	
3	
4	
5	
6	
7	0.076
8	0.136

Notice the factor is called D_3. Just as all the other factors got smaller as the subgroup size got bigger, the A_2 also gets smaller as the subgroup size gets larger. Notice that for subgroup sizes of six or less there is no D_3 constant because negative variability is impossible to ever have.

The upper control limit is computed by adding three standard errors to the average. This interval is approximated by multiplying the R_{Ave} by the factor called D_4. The table below is used to find the appropriate D_4 value. First the number of samples in the subgroup must be known. This number is found in the first column labeled *Subgroup Size* and the D_4 value is then obtained from the first column to the right. This same methodology can be used to compute the lower control limits.

The following calculations are used to derive the data plots for the **R Chart**:

$$R = | X_{Highest} - X_{Lowest} |$$

$$\bar{R} = \frac{R_1 + R_2 + R_3 + \ldots + R_K}{K}$$ where K equals the number of sub-groups

$UCL_R = (D_4)(\bar{R})$ where $(D_4)(\bar{R})$ is an approximation of 3 standard errors above the center.

$LCL_R = (D_3)(\bar{R})$ where $(D_3)(\bar{R})$ is an approximation of 3 standard errors below the center.

D_3 is the conversion factor to arrive at three (3) standard errors for use in determining LCL of R chart.

D_4 is the conversion factor to arrive at three (3) standard errors for use in determining UCL of R chart.

Subgroup Size	A_2	d_2	D_3	D_4
2	1.880	1.128		3.267
3	1.023	1.693		2.574
4	0.729	2.059		2.282
5	0.577	2.326		2.114
6	0.483	2.534		2.004
7	0.419	2.704	0.076	1.924
8	0.373	2.847	0.136	1.864
9	0.337	2.970	0.184	1.816
10	0.308	3.078	0.223	1.777
11	0.285	3.173	0.256	1.744
12	0.266	3.258	0.283	1.717
13	0.249	3.336	0.307	1.693
14	0.235	3.407	0.328	1.672
15	0.223	3.472	0.347	1.653

XBAR and R CHART Conversion

For a subgroup sample size example of five, the D_4 value is 2.114 and the D_3 value is blank. These values are then inserted in the equations above.

$$UCL_{AVE} = \bar{\bar{X}} + A_2 \bar{R}$$

Now let's move back to our simple example, compute all the alarms, and plot the chart.

Explanation R Example

Center Point The R values for each subgroup are calculated and these values are used to compute the control limits. The next formula is used to compute the average R called $\bar{R}$ which becomes our first alarm on the R chart.

$$\bar{R} = \frac{R_1 + R_2 + R_3 + \ldots + R_K}{K}$$

$$= \frac{4+6+1+2}{4}$$

$$= \frac{13}{4}$$

$$= 3.25$$

This $\bar{R}$ is used to approximate and compute the interval around the center. The factor that was derived in the previous section will be used to compute the interval of three standard errors above and below the center. The subgroup size for the example is five, the D_4 value is 2.114, and the D_3 value is blank. This value is then inserted in the equations below.

$$UCL_R = D_4 \ \bar{R}$$

$$= 2.114 \ * \ 3.25$$

$$= 6.87$$

The lower control limit is left blank.

Subgroup Size	A_2	d_2	D_3	D_4
2	1.880	1.128		3.267
3	1.023	1.693		2.574
4	0.729	2.059		2.282
5	**0.577**	**2.326**		**2.114**
6	0.483	2.534		2.004
7	0.419	2.704	0.076	1.924
8	0.373	2.847	0.136	1.864
9	0.337	2.970	0.184	1.816
10	0.308	3.078	0.223	1.777
11	0.285	3.173	0.256	1.744
12	0.266	3.258	0.283	1.717
13	0.249	3.336	0.307	1.693
14	0.235	3.407	0.328	1.672
15	0.223	3.472	0.347	1.653

XBAR and R CHART Conversion

As you did earlier, plot the results on the graph below and compare your results to the graphs on the next page.

R Chart Example

SUBGROUP	1:00	2:00	3:00	4:00	5:00	6:00
Samples	5	5	5	5		
$\bar{X}$ = *Total / Samples*	10.6	10.2	9.8	10		
R	4	6	1	2		

$$\bar{R} = \frac{4 + 6 + 1 + 2}{4} = \frac{13}{4} = 3.25$$

$$UCL_R = D_4 \bar{R} = (2.114)(3.25) = 6.87$$

$$LCL_R = D_3 \bar{R} = (0)(3.25) = 0$$

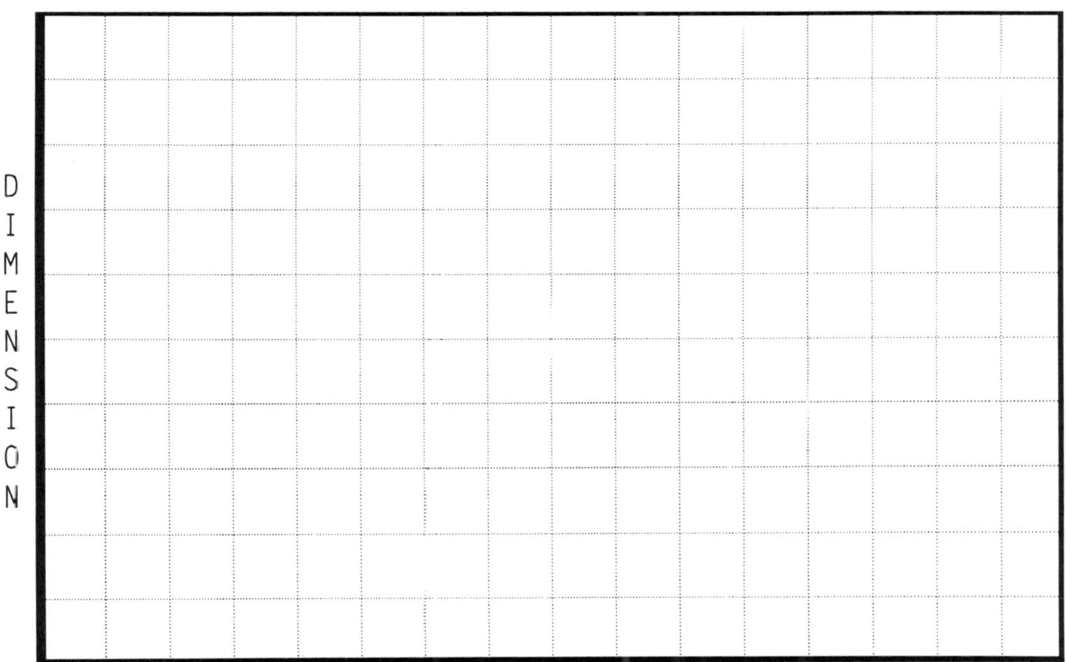

$\bar{x}$ And R Charts

$\overline{X}$ Chart is designed to monitor the movement or changes in the **central tendency** of the process. The **$\overline{X}$ Chart** from our simple example on the prior page is shown below.

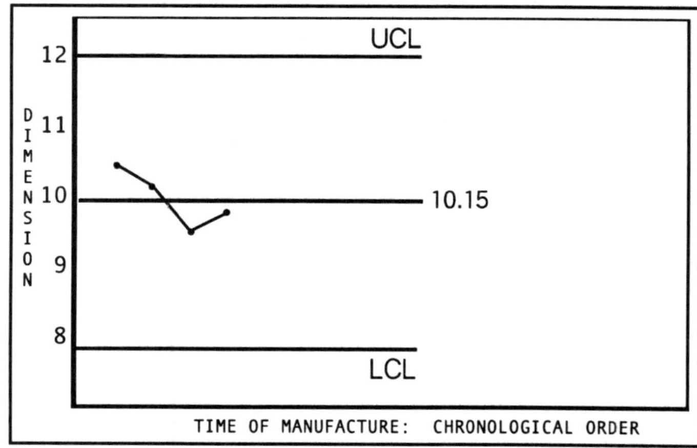

X Bar Chart

R Chart is designed to monitor the movement or changes in the variability of the process. The **R Chart** from our simple example on the prior page is shown below.

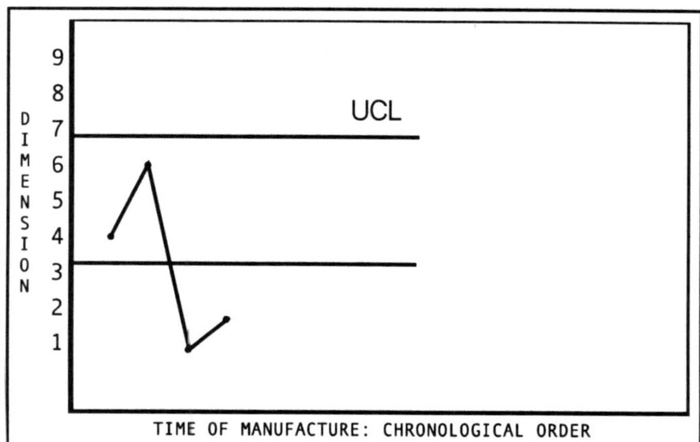

R CHART

CAUTION! Once the process is in control, stop recalculating the control limits. The recalculation of control limits is a joint decision of both strategic and tactical decision makers. Never let computers automatically recompute your control limits without direct human decisions.

$\bar{X}$ And R Chart Exercise

The Candy Plant Revisited

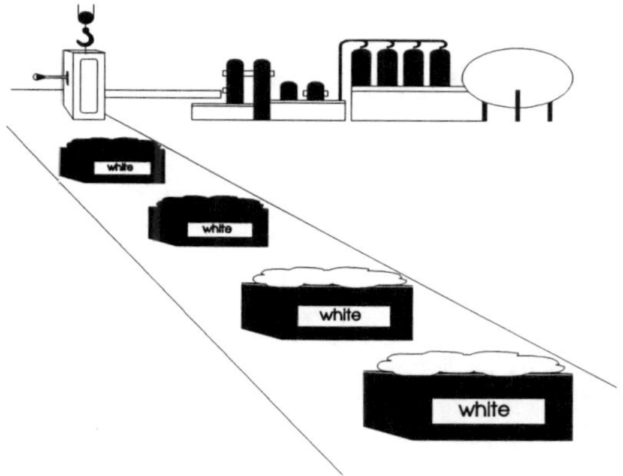

For our exercise, the candy plant is revisited and this time the metric that we are monitoring is the weight of our candy. We want this to be 50 grams. In the table below, readings are taken every hour. The hourly readings are pulled from the conveyor belt in a subgroup of five chocolate candy clusters. These individual chocolate clusters are pulled from the conveyor in consecutive order with the objective of better understanding the weight of the candy as it is produced. The 1:00 to 10:00 readings are shown below.

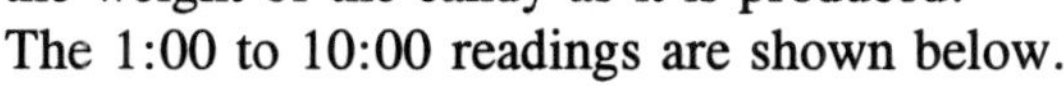

On the following pages, we reinforce the control chart construction. In the example, we will assure that you can construct $\bar{X}$ and R control charts. We are back at the candy plant making chocolate clusters that should weigh 50 grams.

1:00	2:00	3:00	4:00	5:00	6:00	7:00	8:00	9:00	10:00
50	55	60	45	40	50	55	60	45	40
51	56	61	46	41	51	56	61	46	41
52	57	62	47	42	52	57	62	47	42
49	54	59	44	39	49	54	59	44	39
48	53	58	43	38	48	53	58	43	38

Exercise for $\bar{X}$ Control Chart The $\bar{X}$ control chart detects shifts in central tendency. Does our data have any shifts? To answer this question, we must build a control chart. Therefore, the following exercise is designed to complete the control chart that will answer the question. To expedite this exercise, I have already calculated some of the intermediate steps leaving you to compute the control limits. This table with those values is shown below.

	1:00	2:00	3:00	4:00	5:00	6:00	7:00	8:00	9:00	10:00
	50	55	60	45	40	50	55	60	45	40
	51	56	61	46	41	51	56	61	46	41
	52	57	62	47	42	52	57	62	47	42
	49	54	59	44	39	49	54	59	44	39
	48	53	58	43	38	48	53	58	43	38
$\bar{X}$	50	55	60	45	40	50	55	60	45	40
R	4	4	4	4	4	4	4	4	4	4

$\bar{\bar{X}} = 50.0$ $\bar{R} = 4$

Based on the information we have already learned and the data above, calculate the control limits of the chart and then construct the chart on the accompanying form.

UCL =

LCL =

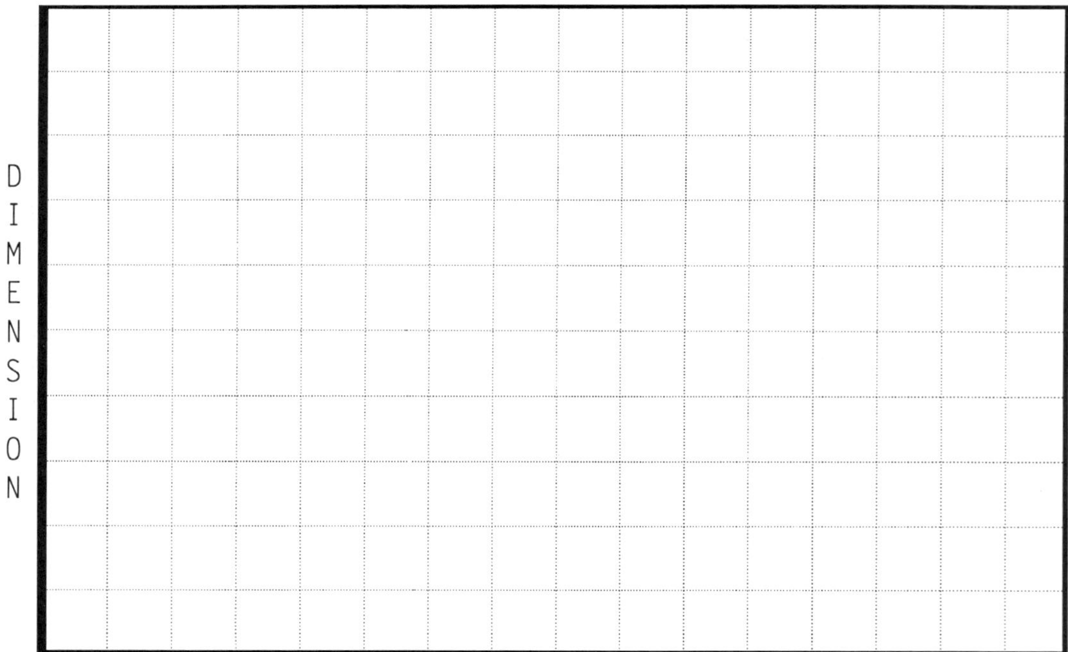

Exercise for R Control Chart The R control chart detects shifts in variability. Now does our data have any of those shifts? Again, we must build a control chart. The following exercise is designed to complete the control chart that will answer that question. I have already calculated some of the R values. Use the table below to compute and draw the R chart.

	1:00	2:00	3:00	4:00	5:00	6:00	7:00	8:00	9:00	10:00
	50	55	60	45	40	50	55	60	45	40
	51	56	61	46	41	51	56	61	46	41
	52	57	62	47	42	52	57	62	47	42
	49	54	59	44	39	49	54	59	44	39
	48	53	58	43	38	48	53	58	43	38
$\bar{X}$	50	55	60	45	40	50	55	60	45	40
R	4	4	4	4	4	4	4	4	4	4

$\bar{\bar{X}} = 50.0$ $\bar{R} = 4$

Use the information above to calculate the control limits for R chart and then construct the R chart in the form below.

UCL =

LCL =

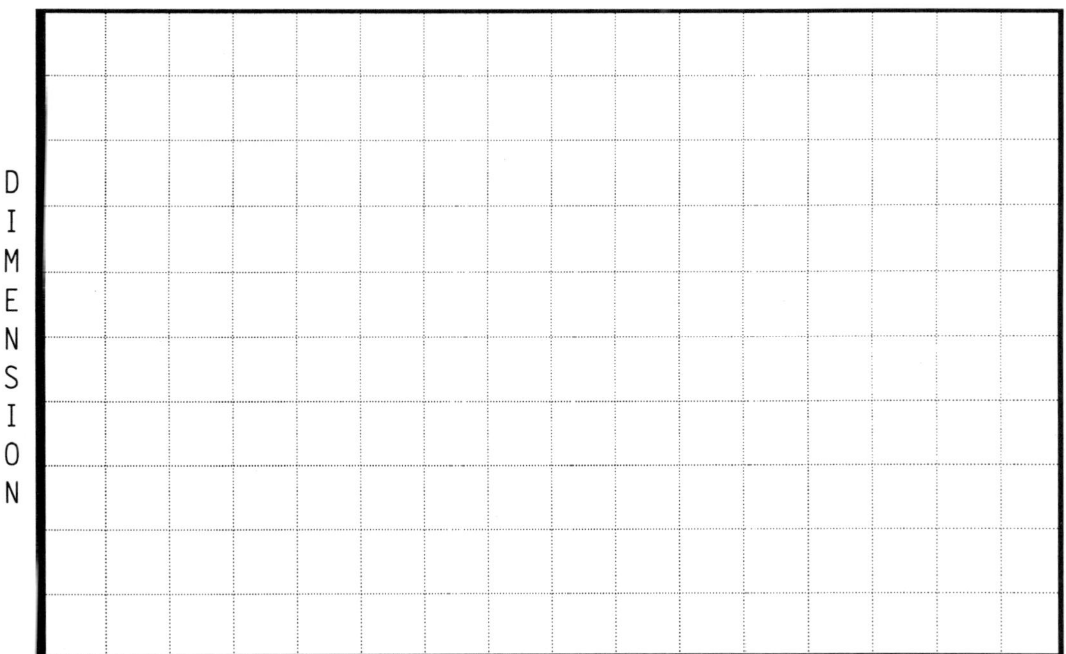

Between and Within Group Shifts

Many people are surprised that the posted results of the computed values do not provide charts that are in control. Our exercise shows this situation. The charts below could be a broader example of the situation encountered with our exercise.

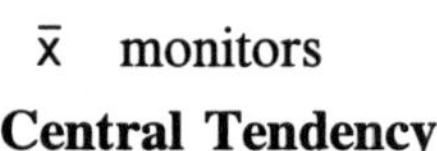

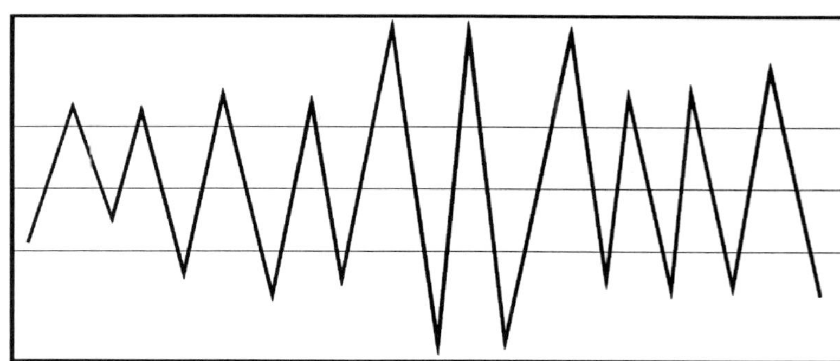

R monitors
Variability

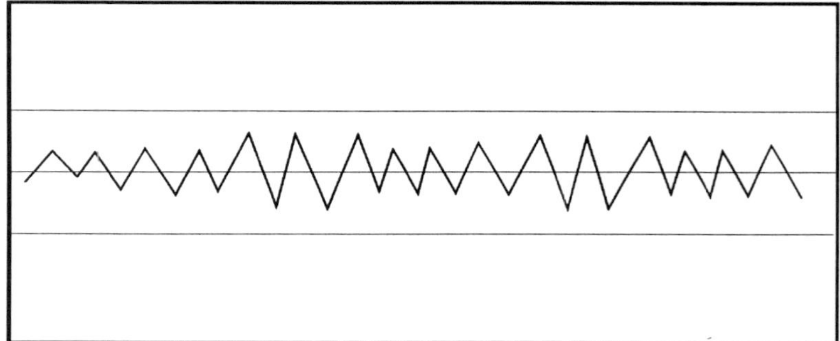

Not only does the $\bar{X}$ Chart monitor shifts, patterns, and trends associated with central tendency, but the $\bar{X}$ also monitors *between* group shifts. The R Chart monitors shifts, patterns, and trends associated with variability. The R chart also monitors part to part variation; the R Chart monitors *within* group shifts.

How can the $\bar{X}$ chart be so erratic when the control limits are from the data? The alarms of a control chart are computed from R values. The R is the least variability that we would expect from a machine because the R values are calculated from consecutive reading in the subgroup. Since we normally pull five units, the variability that we use to compute the alarms should be very low.

Often charts that look like the above are examples of violations of the second principle of process management. Rather than run the process correctly and consistently through the use of control charts, operators make opinion-driven adjustments that are not needed. This is symptomatic of something happening in between the subgroup readings. I would immediately begin to take more frequent readings to understand what is causing these changes.

Between Group Shifts The $\bar{X}$ control chart not only monitors the central tendency but also can detect shifts from group to group. The mechanics of using the part to part variability to compute the control limits provides this additional information.

$\bar{X}$ → **Detects → Central Tendency or GROUP to GROUP shifts.**

Within Group Shifts The R chart in addition to its main purpose of monitoring variability is also checking for any shifts in part to part change.

R → Detects → Variability → PART to PART change

This is the tip of the iceberg for control charts. Now we must summarize what we have learned about the construction of a control chart and move on to understanding how to interpret and use this tool.

Chapter Summary

Remember the table below is used to compute our control limits and is based on the use of the number of samples in the subgroup.

A_2 is the conversion factor to arrive at three (3) standard errors for use in determining UCL & LCL of $\bar{X}$ chart.

D_3 is the conversion factor to arrive at three (3) standard errors for use in determining LCL of R chart.

D_4 is the conversion factor to arrive at three (3) standard errors for use in determining UCL of R chart.

Subgroup Size	A_2	d_2	D_3	D_4
2	1.880	1.128		3.267
3	1.023	1.693		2.574
4	0.729	2.059		2.282
5	0.577	2.326		2.114
6	0.483	2.534		2.004
7	0.419	2.704	0.076	1.924
8	0.373	2.847	0.136	1.864
9	0.337	2.970	0.184	1.816
10	0.308	3.078	0.223	1.777
11	0.285	3.173	0.256	1.744
12	0.266	3.258	0.283	1.717
13	0.249	3.336	0.307	1.693
14	0.235	3.407	0.328	1.672
15	0.223	3.472	0.347	1.653

XBAR and R CHART Conversion

$\bar{X}$ Chart Control Limit Calculations

$$UCL_{AVE} = \bar{\bar{X}} + A_2 \ \bar{R}$$

$$LCL_{AVE} = \bar{\bar{X}} - A_2 \ \bar{R}$$

R Chart Control Limit Calculations

$$\mathbf{UCL_R = (D_4)\ (\bar{R})}$$

$$\mathbf{LCL_R = (D_3)\ (\bar{R})}$$

$\bar{X}$ → Detects → Central Tendency or GROUP to GROUP shifts.

R → Detects → Variability → PART to PART change

Remember all control charts are tactical tools to detect change in the product and process. A control chart can never tell you how well the product meets the customer's expectation. This will come in Chapter 10: Capability when we cover the strategic tools for management and customers.

Chapter Seven: Using a Control Chart

Chapter Seven will cover the interpretation, reading, and use of control charts. Since control charts are based on a normal distribution, coverage of the $\bar{X}$ and R interpretation should make all the other charts' use and reading very simple. All control charts use the same set of alarms to detect change.

The overall objective of control chart methods is to detect product or process changes. The term "**out of control**" simply means a change has been detected. The demonstrated effectiveness of a control chart is attributable to its objective and systematic analytical methods. *Shortcuts and "make-do" procedures lead to invalid conclusions and costly mistakes.*

In Chapter Six, all the methods to effectively control a process and product were discussed. Let's review the following steps:

- Calculate initial control limits
- Establish that the process is in control
- Continually survey the control charts for alarms (out-of-control conditions)
- Assess the capability of the in-control process

We covered the initial control limit calculations in Chapter Six. Different control alarms for each machine would not be surprising and as a matter of fact are expected. We also explained the mechanics and calculations for control limits and discussed how to draw the control chart. Now that we have the control chart, we must learn to use it effectively. This chapter deals with the steps of getting the process in control and then assuring a continuous surveillance program to detect any changes.

Chapter Seven will cover:

Establish that the process is in control As each out-of-control condition is identified and dealt with, this procedure is repeated until no alarms have been detected. When the process is in control (no alarms have been detected), the control limits are locked in place. We must now explain the reading and interpretation of the control charts.

Continually survey of the control charts for alarms (out-of-control conditions) Data is gathered to monitor against the alarm mechanism (control limits) for changes in the conditions of the machine. When an out-of-control or change is detected, the chart has done its job. The following tasks must be undertaken.

i. Investigate for the cause
ii. Alert management to the change
iii. Decide what to do

This chapter will explain the details of the tasks associated with dealing with an alarm: both what must be done and who should do it.

How are Control Charts used?
The flow diagram on the next page shows the general steps of a control methodology. The control chart component is one player on the SPC team. A control chart helps the operator understand his machine and establish how to correctly and consistently run it. The chart also establishes the typical patterns for the machine and alerts the operator to any deviations from the norm. These deviations are change (out-of-control).

Remember, control charts are only one player on the team. We must have all the players in place for SPC to be effective. The flow shows that an ongoing, never- ending data gathering effort is maintained. If no change is detected, we must wait for the next readings. If a change is detected, our investigation begins to determine the cause and assess the impact. Once all the prior steps are complete, then we decide what action to take.

Detect Change
Control charts are a tactical tool to detect change in a continuous operation. A series of easy to use and read alarms must be developed for the tactician (operator). These alarms must be simple but effective.

Chapter Two covered the math trick that forced a normal distribution and makes the alarms simple because no decision is required by the operator as to the type of distributions. All our alarms are based on the normal distribution.

This chapter will discuss in detail the interpretation rules of control charts and how they were derived.

Once a change is detected, we are just at the start of our journey however; we must move on to our investigation efforts.

Initiate Investigation
Our investigation is comprised of two parts: one tactical and one strategic. A clear understanding of the roles of both the strategic and tactical players is key. Once the individual roles are clear, the players must function together as a team for effective action.

Tactical decision makers (operators) begin investigating to determine the cause of the alarm. These causes will increase our understanding of the process and product. This knowledge is crucial for improvement. Once the knowledge has been learned, it must be shared and retained for all associates. The tacticians must also alert the strategic decision-makers that a changed has occurred so that an assessment of the impact of the decision can be made.

Coming in Chapter Ten The strategic decision-makers must simultaneously (while the cause is being investigated) assess the impact of product or process change on the customer.

Once the cause has been identified and the capability determined, a course of action must be chosen. This is a team decision by both the strategic and tactical decision-makers.

When is a Change Real???

Often the natural ebb and flow of a process is mistaken for real change. To prevent that misinterpretation, a set of benchmarks are required to establish what is typical for a particular machine. The control chart's limits provide the means to differentiate between what is typical and what is unusual behavior for a process and product. Remember, what is usual for one machine may be bizarre for another. SPC provides the tools to separate typical from unusual.

The SPC tool of control charts is comprised of a few simple alarms that note a change is taking place and that an immediate initiation for the causal search should be advanced at full speed. The next pages describe the common patterns found in control charts, and where to begin looking for causes. The chart below shows the steps for the control phase of SPC.

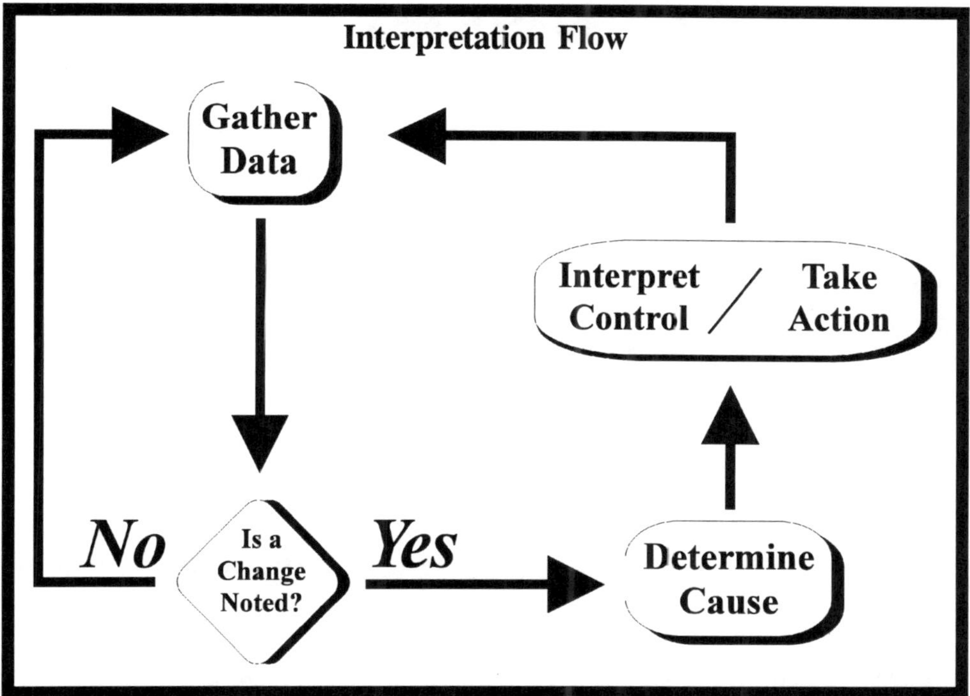

The above chart depicts the steps to action based on a control chart alarm.

Remember, never take action just because an alarm is noted. The cause must always be found before taking action.

STEPS TO INTERPRETATION for an Alarm

From our earlier discussion of basic statistics, we must be aware of both variability and central tendency. A chart is provided to monitor both variability and central tendency. Variability is so often skipped, misunderstood, and ignored. Thus, I always monitor the variability chart first and then go to the central tendency chart second. Once each component has been monitored for change, then I look at both charts together. To summarize: the following steps are followed:

1. Check Variability
2. Check Central Tendency
3. Check Both

How Did Control Chart Alarms Originate? The alarms of a control chart are based on the area under the curve which represents the probability of occurrence.

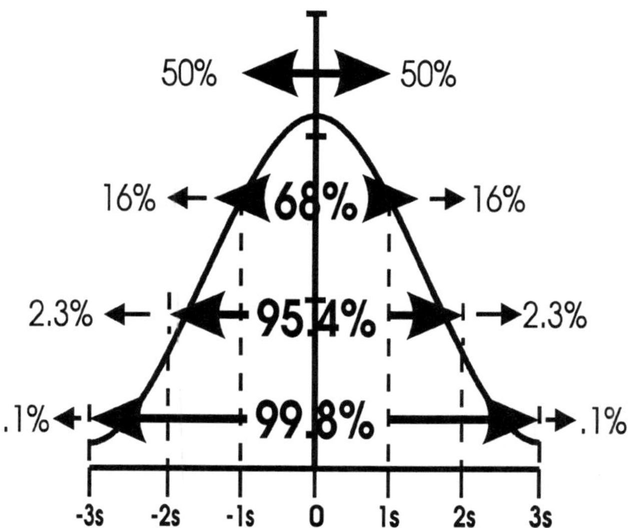

An occurrence outside plus or minus one standard deviation (one is the z-Value) equates to a probability of 16 percent above or 16 percent below or 68 percent inside. An occurrence outside plus or minus two standard deviations (two is the z-Value) equates to a probability of 2.3 percent above or 2.3 percent below or 95.4 percent inside. An occurrence outside plus or minus three standard deviations (three is the z-Value) equates to a probability of 0.1 percent above or 0.1 percent below or 99.8 percent inside.

Remember, our control limits were defined as being three standard errors away from the mean. The control limits are set such that the odds of having a value outside either of the limits is 0.1 percent or, stated very simply, the odds are very remote that a point will be outside the limits without truly being a change. Since we are very sure that a change is taking place, then an immediate investigation as to the cause is initiated.

Since the odds of being outside the control limits are remote, then change and the search for the cause is seldom a wild goose chase. This rarity of the alarm is the reason that the other alarms are all based on about the same probability. The next section of this chapter covers the basic set of control chart alarms and how they were each developed.

Single Point Outside Control Limits

Any data point outside control limits requires investigation, because the odds that a controlled process could produce such a measurement purely by accident are very remote. This rule applies to both the $\bar{X}$ or R control chart.

Any subgroup average falling outside the $\bar{X}$ chart control limits is an indication that the central tendency is changing. An immediate investigation is required.

Remember, as we stated before, never act on the alarm. Before any action can be taken, the cause of the alarm must be identified. Later in this chapter we will discuss causes. The control chart does not tell us the cause but only alerts us to start the investigation.

Acting on the alarm would be like immediately throwing water on a kitchen fire after a smoke detector sounds only to find out later that the fire is a grease fire and water causes a grease fire to spread.

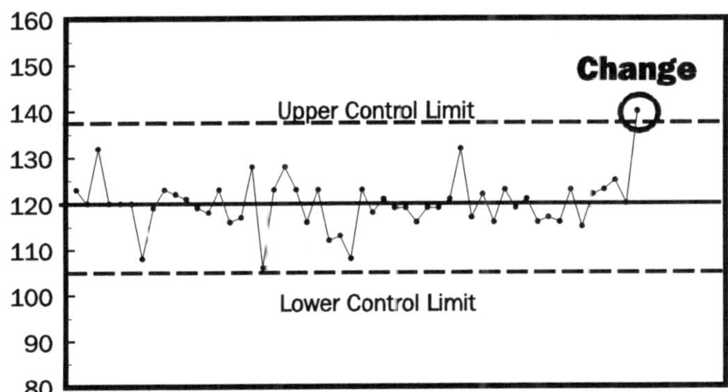

The chart above should trigger an alert to investigate and determine the cause and the impact.

A Series of Points Outside Control Limits

Since the odds of a single data point outside the control limits is 0.1 percent, then a series of data points outside must really be rare. The odds of having one point outside the control limits is 0.1 percent or 0.001.

The second occurrence of having a point outside the control limits is also 0.1 percent or 0.001. The odds though of having two **consecutive** points outside the control limits is the first probability multiplied by the second probability. Thus the probability of having two consecutive outside the control limits is 0.001 times 0.001 which is 0.000001 or 0.0001 percent.

Number of points outside the control limits	Probability of Occurrence
1	.1%
2	.0001%
3	.0000001%
4	.0000000001%

As is evident in the table above, the probability of having multiple consecutive data points outside the control limits becomes very rare. Four straight data points outside the control limits which do not reflect a change would almost be impossible. This leads us to the conclusion that this must truly be a shift in the metric that is being monitored. The control chart below shows this particular situation.

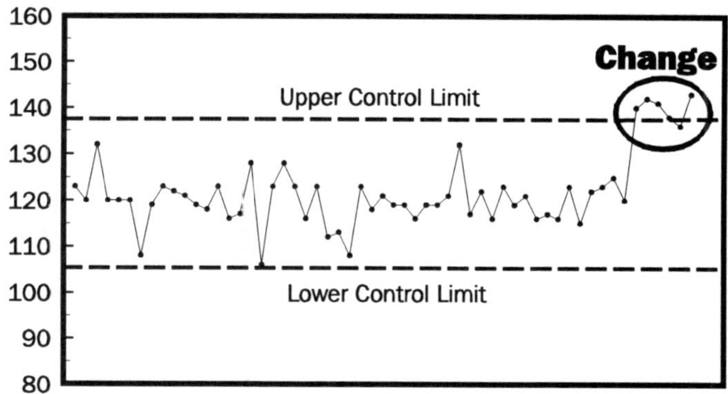

All our alarms apply to control charts for both central tendency and variability because the alarms are based on the probabilities of a normal distribution.

Sudden Jump in Level
Many people mistakenly only look for the alarms that we just covered. Control charts were designed to detect much more subtle changes than those. Because control chart methods are designed to detect minor change, many of our alarms are based around values that are inside the control limits. These alarms are geared to detect the most minor of shifts that are outside the typical for any one particular machine. The first change that is detected with data inside the control limits is a **jump** in data.

Seven Consecutive Data Points on One Side of The Center

This particular alarm was geared for minor shifts above the mean of the central tendency or variability or minor shifts below the center.

The alarm is seven consecutive data points all above the center point of the control chart, but none of the data points are outside the control limits. The chart below reflects that particular alarm.

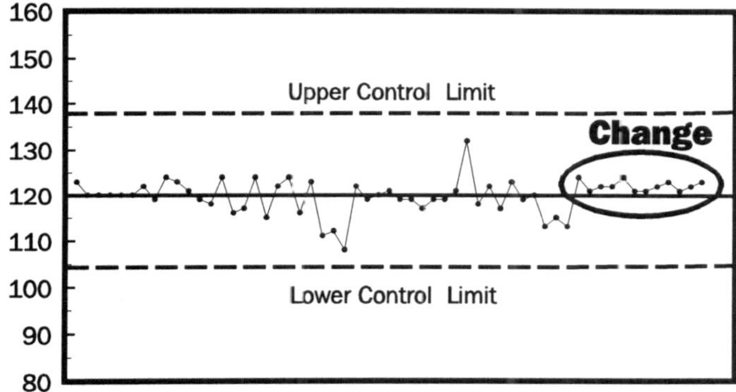

The chart above shows a series of data points all above the center line. The more consecutive the values on one side of the center, the bigger the alarm. The first indication of an alarm is when seven consecutive points are on the same side of the center line.

A similar alarm to the above is seven consecutive data points all below the center point of the control chart, but none of the data points are outside the control limits. Restated, the alarm is seven consecutive values on one side of the center of the control chart with none of the values falling outside the control limits.

All the alarms are alerts that a change has occurred in the process or product. Never stop the machine because of an alarm on the control chart. This alarm may be a signal of an improvement!

Why Use Seven Consecutive Points as the Alarm?

The normal distribution has fifty percent of the values above and the remaining fifty percent of the values below the average. This probability would lead us to expect many values above the center or any one value by itself is fifty percent.

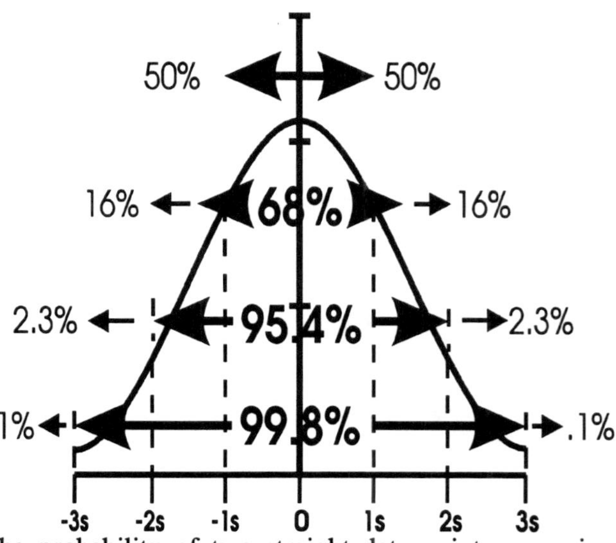

The probability of consecutive values outside the control limits was covered earlier. When consecutive occurrences happen, the probability is multiplicative. The probability of having two consecutive data points above the center is calculated by multiplying together the probability of occurence of the individual data points.

One data point above the center is itself a 0.5 probability. Therefore, the probability of two straight data points occurring consecutively is 0.5 multiplied by 0.5 or a 0.25 probability.

The table below shows the probability of having seven straight values on one and only one side of the center and no values outside the control limits.

Consecutive Points on one side of the Center	Probability Calculations	Probability
1	.5	.5
2	.5*.5	.25
3	.5*.5*.5	.125
4	$.5^4$	.06
5	$.5^5$	.03
6	$.5^6$	.016
7	$.5^7$	.008

The seven consecutive points on one side of the center (either seven above the center or seven below the center) applies for both the charts monitoring the central tendency and variability.

The seven rule is used because the probability is so close to the same probability as one point outside the control limits. This keeps us using the same basic logic for all our alarm conditions.

Trend

Another minor shift that is important is a **trend**. Trends can take the form of going up or coming down. Once again minor shifts are the objective of the control chart's detection mechanism. This trend is also found where the data is inside the control limits.

Seven Consecutive Decreasing (or Increasing) Data Points

This particular alarm, illustrated below, is geared to detect a slow shift in a downward direction ("trending down"). It is described as seven consecutive points with each point lower than the last point and no points outside the control limits.

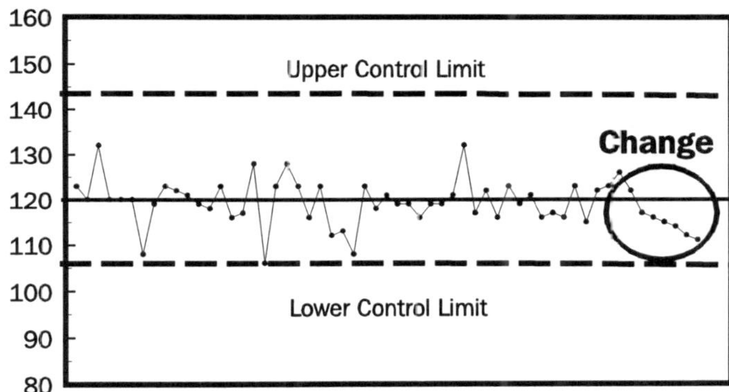

The same idea applies to the reverse, where the trend is increasing. This particular alarm is geared to detect a slow shift in an upward direction ("Trending up"). This alarm is seven consecutive points with each point higher than the last point and none of the points outside the control limits.

Remember, the normal distribution is symmetrical. The symmetry of the normal distribution explains how this alarm was developed and why we use seven points. The probability calculation for this alarm is similar to the alarm of seven points on one side of the center.

Control charts must be viewed in a total context of all the information that is available. Just because no control chart alarms have gone off does not mean that other information might not cause an alert. As an example, if a drill bit is dull (needs replacing) the information about the bit should tell us to change the bit even though no control chart alarms have sounded.

Too Close

Revisiting the probabilities of a normal distribution, we find that 68% of the values should fall inside plus or minus one standard deviation. For our application, 68% should also fall inside one standard error. Based on our understanding of a normal distribution then, approximately 68% of the data points should be within one standard deviation of the average. Many people refer to the area inside plus and minus one standard deviation as the **zone one** area.

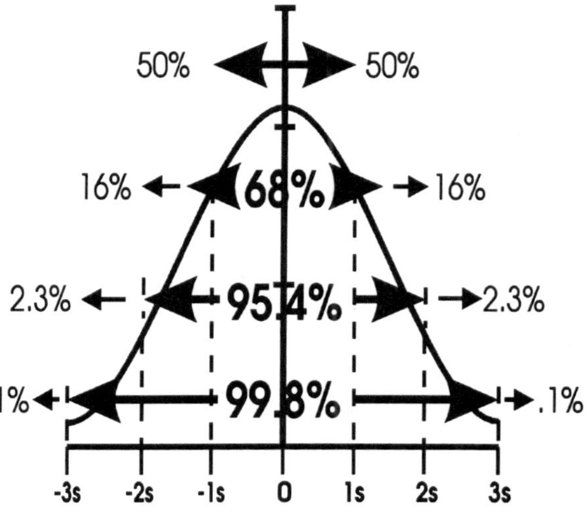

More Than 68% of the Data Within Zone One

If too many of the values are clustered around the mean, an alarm is indicated. Remember this is not an indication of bad product or good product, but an indication of change that needs to be investigated.

On the control chart below, all the data points are inside the zone one area (plus and minus one standard deviation). This chart is an example of the "too close" alarm going off. This alarm like all the other alarms must be investigated immediately.

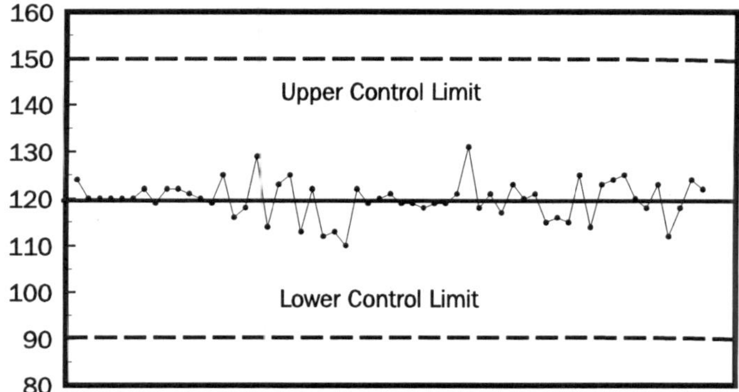

Minor change going unnoticed over time may seem trivial, but these minor changes have a compounding affect and over time allow us to miss major improvement opportunities or slide into major defect problems.

Too Far Apart

Still viewing a normal distribution, the number of data points above one standard deviation but with no points outside the control limits should be 16% of the data points. The number of data points below one standard deviation with none outside the lower control limits is also another 16%. The total data points within the control limits and outside zone one is 32 percent.

The area is comprised of the area from one standard error to two standard errors is referred to as **zone two**. The area from two standard errors to the control limits is called **zone three**. The total number of data points in zones two and three is 32 percent.

The alarm here is if more than 32 percent of the data points are in zones two and three. This becomes just as big an alarm as all the prior alarms.

More than a Total of 32% of the Data in Zones 2 and 3

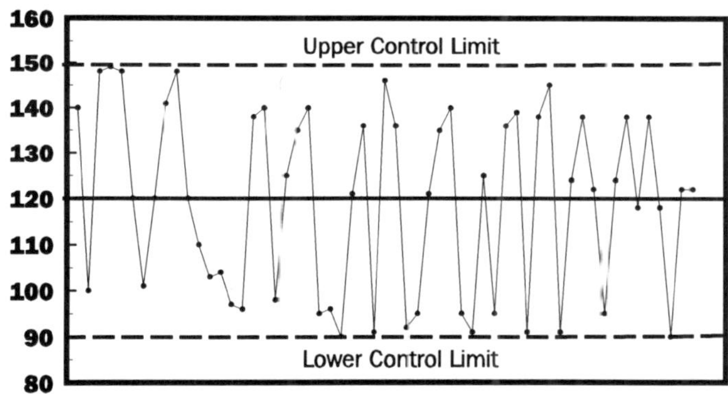

On the control chart above, none of the values fall inside zone one, but no values are outside the control limits. This is an example of the too close alarm.

This particular alarm is symptomatic of a bigger problem. A single control chart might be used (inappropriately) to monitor more than one machine. Because each machine has its own distribution, when these unique machines' distributions are laid on top of one another, a false distribution results.

The bigger problem might be formed with the individual distribution of machine one and machine two shown below.

Candy Machine One

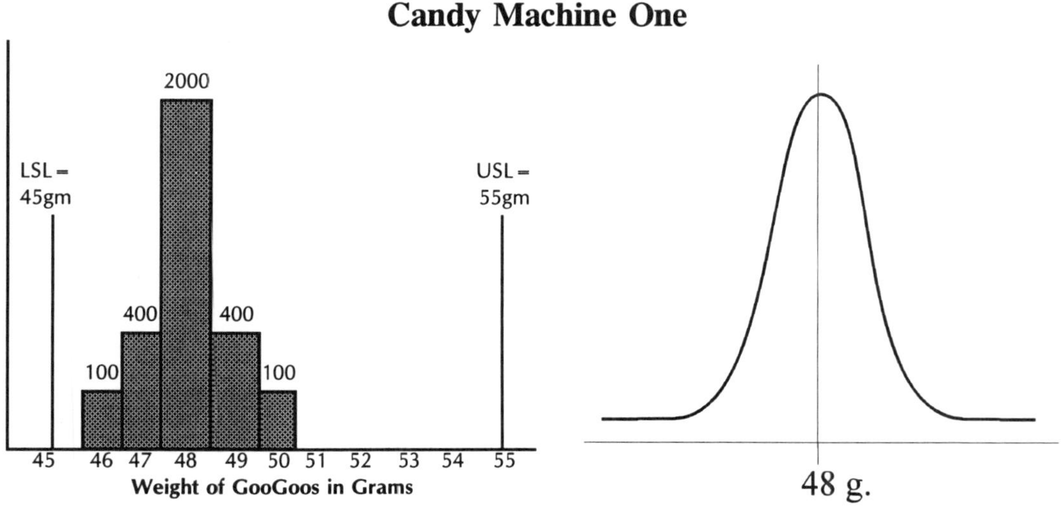

Candy machine one's distribution is centered with an average of 48 grams as shown in the above graphs.

Candy Machine Two

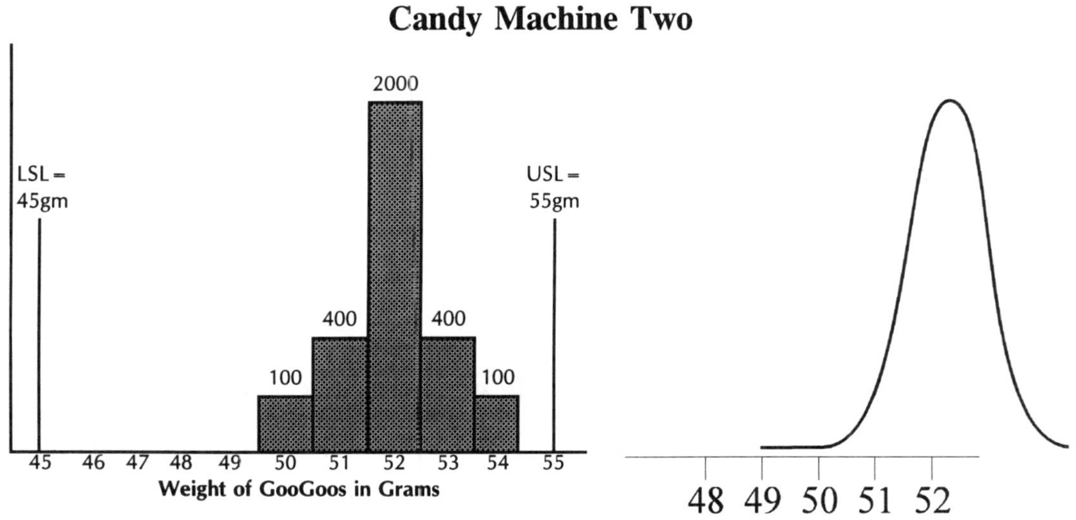

Candy machine two's distribution is centered with an average of 52 grams as shown in the above graphs.

Care must be taken to assure proper grouping and to avoid the mixing of the two machines production.

When the two machines' metric measurements are lumped together, an improper distribution is formed.

The two candy machines' distributions looks like the graph to the right.

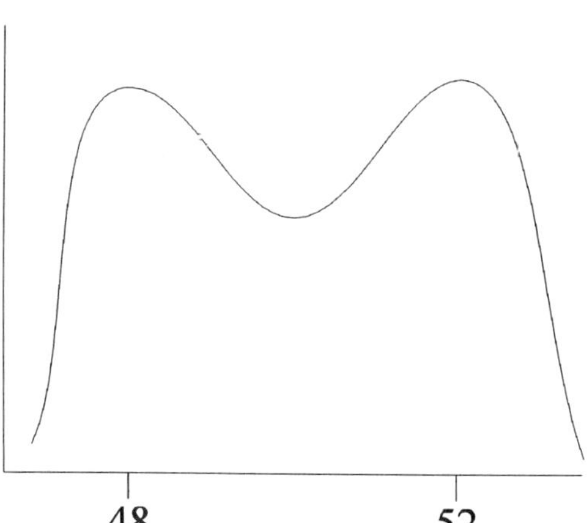

Are We Looking at One Process or Many? A controlled process behaves as a single system, and over the long term, the sample averages drawn from it will form a normal distribution, driven by a single system of common causes.

If the measurements are actually being drawn from more than one process, the distribution of sample averages will not be normal, but lopsided or even bimodal (with 2 peak values). This could happen if the output of two workers or machines were combined on the same chart.

The histogram below is representative of how a bimodal distribution (two machines' distributions) might look. Notice the two spikes that are caused from the two different distributions.

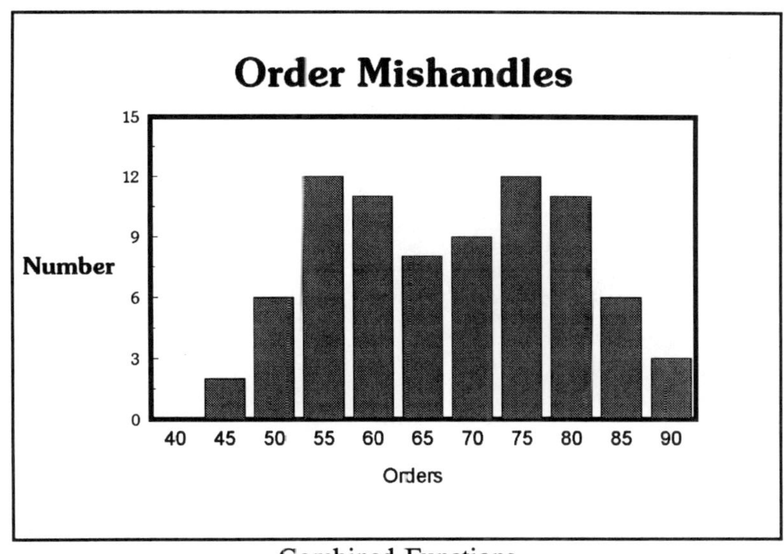

Combined Functions

Recurring Cycles

Any recurring pattern that deviates from the normal distribution must be viewed as an alarm. This alarm is not as specific as all the others. Since control charts are based on the probabilities of the normal distribution, the 68% within one standard error, 95% within two standard error, and 99.8% within two standard deviations must be used to search for these patterns. As an example, when a pattern of values of values outside the two standard error area is detected an alarm to start the investigation must be initiated.

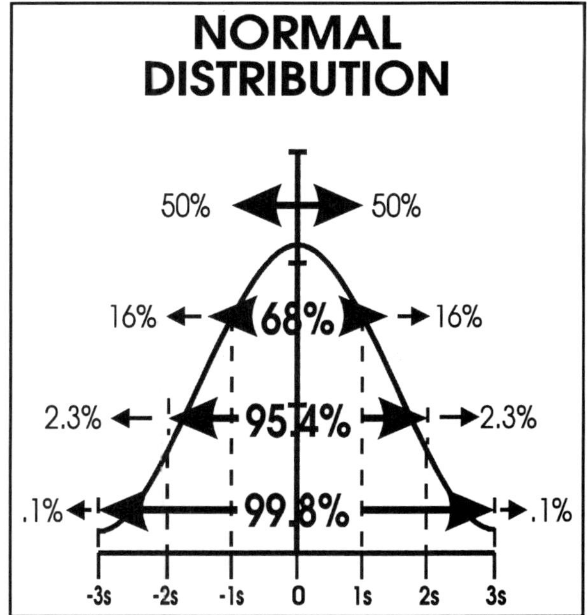

When any pattern or cycle violates the normal distribution, an alarm is triggered. These alarms are just as critical and important as the all the standard alarms.

Summary Of Interpretation Guidelines

The table below summarizes the alarms that we use to monitor or detect any shifts, patterns, or trends.

- Single point outside the control limits
- A series of points outside the control limits
- Seven consecutive points increasing (or decreasing) from the last point
- Seven consecutive points above (or below) the average
- More than 68% of the points in the first zone
- More than 32% of the points outside the first zone

Alarm Zones

Zone 1	
Average + 1 Std_{Ave}	68% of the points fall inside
Average - 1 Std_{Ave}	
Zone 2	
Average + 2 Std_{Ave}	95% of the points fall inside
Average - 2 Std_{Ave}	
Zone 3	
Average + 3 Std_{Ave}	99.8% of the points fall inside
Average - 3 Std_{Ave}	

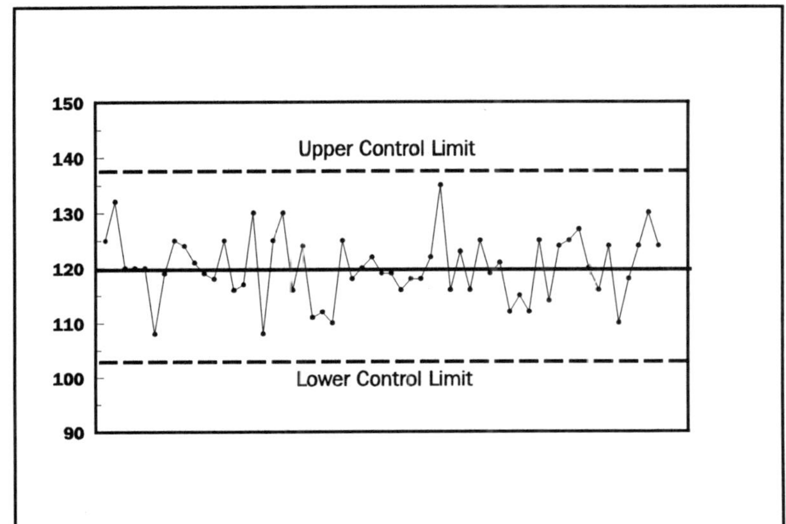

What is a Control Chart Alarm? The control chart alarm is a change detection mechanism. These process or product changes are outside the typical behavior of the machine. Remember, change is not always bad but can be either an improvement or a deterioration. For progress (improvement) to be made, change has to occur.

The control chart is the tactical tool to identify when the change is enough for us to investigate. Our investigation is an attempt to determine the reason the change took place which is referred to as *finding the cause*. As we find the cause or causes, our knowledge will grow and with knowledge comes improvement.

In many ways the control chart (when used in conjunction with strategic tools) is our improvement detector. Actually the control chart can only find change, but without the control chart, change will go unnoticed. This is particularly true of minor change. Over time, minor change will amount to major change.

Now, once a change has been documented as more than just a fluke, the tactical role becomes even more focused on the correct and consistent running of the machine. The tactical operator must now focus on what caused the change while the strategic focus is to assess the impact. What is a cause and how do they work?

What are Causes?

Our Knowledge Base linking is the attempt to determine what caused something to happen. Effects are the things that happened. The deduction of what caused the effect is the most difficult task we will encounter. Since all facts are never known, making the determination of the cause of an effect is often difficult. *The true cause or causes may be quite unknown.* True cause identification is an ongoing effort which requires a vast amount of analytical experience. It will require much time, analysis, thought, and effort. We will be continually adding to our true causes, and in doing so, we will increase our Knowledge Base.

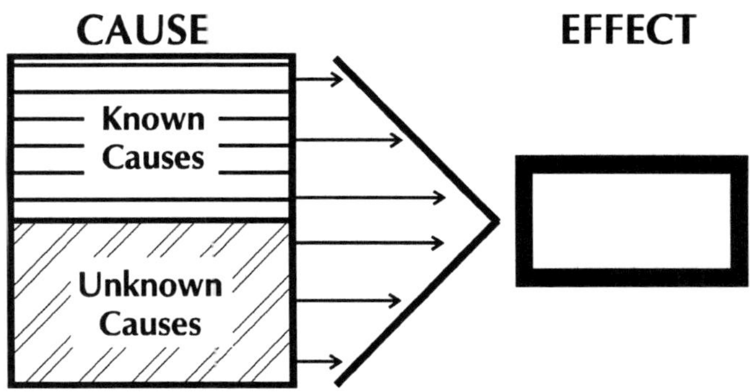

Since we never know all the facts or have a complete picture, we must continually review our determinations and findings. We must never view our determinations as laws or standards that cannot be improved. The disparity of known causes directly impacts our determinations and is due to a lack of all the facts or a failure to understand the meaning of the facts we have. The magnitude of this disparity must continually be investigated.

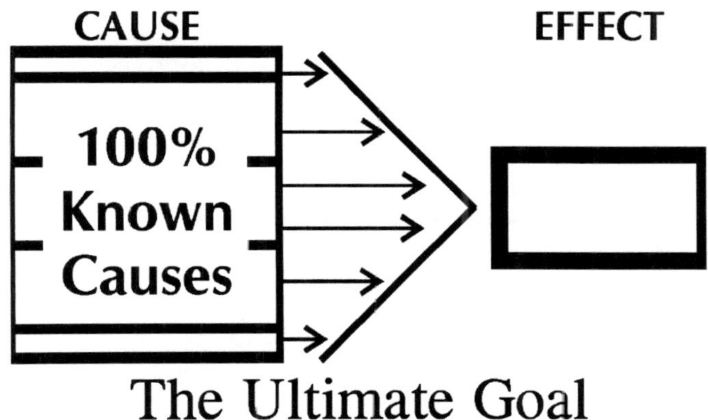

Who is Best Equipped to Find the Cause??

We first think of the machine designer, product designer, or process engineer as the most likely candidate to determine the cause of the change. These individuals are well equipped with technical skills. However, the investigation process must begin quickly so that the causal identification trail does not get cold. The problem with all the people listed is that their timely availability is questionable. Engineers, management, and designers are strategic and must be involved in the final action plan.

Change happens at all times of day, weekends, holidays, and even at 3:00 AM on Saturday. Since our efforts must follow the *Church of What's Happening Now* philosophy, the person closest to the machine must be the causal searcher. This narrows our scope to the prime person most familiar with the machine - the operator. The operator, our tactician, will lead the efforts to finding the cause of the change.

All the technical resources that can support the operator must be brought into play to find the cause. As the operator finds these causes, the knowledge base about the process and product will be increased.

Our objective is to continually reduce this disparity and achieve total knowledge of the *True Causes*. Over time we must strive to increase our Knowledge Base.

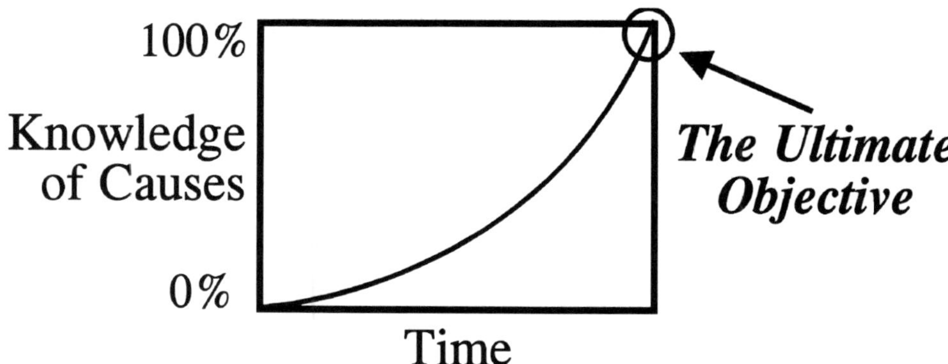

Effects are seldom the result of a simple single cause. Since all causes are seldom known, we should expect gaps in our Knowledge Base. These gaps of knowledge should keep us alert to the fact that effects are not always the result of known causes. Serious trouble will be encountered if the facts are forcibly stretched to explain the effect. A wise person is one who appreciates his lack of total knowledge.

To build our total Knowledge Base, our analysis must be based only on a historical record. The historical records allow us to begin to build a causal linking. This will not preclude our adding knowledge of changes in the future to better understand the cause. This is like driving a car. We look through the rear view mirror to understand where we have been. This view also allows us to see cars coming from the rear. We look through the front windshield to understand things coming up such as a curve in the road or a car on our side of the street.

The data for our analysis must begin at the basic element. Our analysis must begin both from the bottom and from the top allowing the art of causal analysis to take place.

Our causal linking is easy when we are viewing the most immediate occurrence that is known. The further removed the causes are from the effect, the more difficult the linking task. This linking must track the links of the chain from process to product. All products and processes are interrelated. This interrelationship impacts the outcome (effect).

To better understand the investigation and analysis that will be required, let's define the type of causes that can come into play.

Assignable Causes: As we begin to study our information, a clear understanding of what a cause means and the types of causes which are possible is crucial to understanding what actions are required. We must first understand that we are looking for change. Change includes both improvements and deterioration. We must understand which type of change we are dealing with in order to understand our courses of action.

> **Assign** 1. To set aside or give out in portions or shares. 2. Attribute; To consider as resulting from, proper to, or belonging to a person or thing.
>
> **Cause** 1. The producer of an effect, result, or consequence. 2. The one, such as a person, an event, or a condition, that is responsible for an action or a result. 3. What brings about or is associated with an effect or result. *A cause is an agent or condition that permits the occurrence of an effect or necessarily or invariably leads to a result.*

When we put the dictionary's definitions into action, we conclude that assignable causes are those which definitely result in a particular effect. As we read the definitions, we see that all causes do not have equal impact.

As we increase our Knowledge Base, we must strive to include all causes no matter how small their impact. A series of minor causes can lead to a cumulative impact that is very large. Too often we are only looking for the Big Bang.

Causes are neither good nor bad but simply a statement of fact. A cause is why something happened or changed. For us to make improvements, we must first detect change and then determine its cause. We will diligently search for all causes, both those bringing about improvement and those resulting in deterioration. This information will allow us to increase our Knowledge Base of the process.

The reason causes are so important is that they give us the information to know what will happen in a particular situation. Many times an associate who has been on the job for 20 or 30 years may still run to the back end of the machine to find out what has happened every time that he changes a knob. There is major difference between 30 years on the job and *30 years of experience*. With 30 years of experience, we have built a huge Knowledge Base that will tell us what is going to happen to the process.

Types of Assignable Causes: For us to effectively make a decision, to understand its impact, and to know the correct type of decision, we must know what type of cause is involved. There are three kinds of causes: 1) **Special**, 2) **Common**, and 3) **Tampering**.

Special Causes are causes within one's control. The person running the machine is directly involved in the creation of a special cause. Special causes are isolated to one particular machine and impact only that one machine. As I drive down the street, the car's course adjustments are caused by my movement of the steering wheel. The car's speed is determined by my pressure on the accelerator.

In the same way, only an operator on a shop floor can make adjustments to his machines. These could include the speed the machine runs, the temperature of the machine, and the feed of the machine. No matter how the operator might desire these settings to impact other machines, it's pretty obvious that these changes, like the car's course adjustments, will have no effect on other machines.

Since special causes are local in nature (assigned to the driver or the operator), local action is required to resolve them. The operator must resolve the effects by dealing with the cause locally.

Special causes are the *operator's responsibility* for resolution. Local actions are usually required to eliminate these special causes and can usually be taken by people close to the process. Tactical decisions and tactical actions are generally used to deal with special causes. The driver of the car above must make course corrections to assure the safe driving of the vehicle. Such corrections are tactical decisions.

Common Causes are causes outside one's control. These influences generally occur across the board for all processes, thus the name *common.* The operator is usually not the generator of common causes. If the purchasing department buys bad peanuts for our candy, the operator is not involved in this cause of change. Common causes are things that cross the boundaries of one machine into a group. Common causes impact multiple machines, and the impact is similar on all involved. Using the driving examples, I might choose to slow the speed of my car on a rainy day. This adjustment to my speed is caused by the rain which is outside my control. Most other drivers would slow down also due to the bad weather conditions. All the cars are being impacted by the weather, and weather is common to all; no action by the drivers, either singularly or together, could change the weather.

All operators and their machines might be impacted by a shift to a new raw material such as those new peanuts for our GooGoos. The peanut raw material from the new source may be inferior and cause a rash of defective GooGoos. No matter how sincere, conscientious, and diligent the operator, he cannot directly resolve the defects or change the raw material.

Since common causes are outside local control, sources outside the local area must resolve these causes. Generally this is thought of as a managerial issue because management has overall control of the facility and the actions on the system almost always require management action for correction. Strategic decisions and strategic actions are generally used to resolve common causes. Purchasing must resolve our bad peanuts in order for good tasting GooGoos to be manufactured. Purchasing must make strategic decisions to resolve the problem or strategically decide to accept the defective GooGoos. No amount of local action can resolve this problem.

Tampering Causes are changes that are caused by too much control. Tampering occurs when we are trying to take action on something that should not be changed. This is typically caused by our failure to understand and comprehend the *Second Principle of Process Management*.

Second Principle of Process Management Division of Labor is the framework for all aspects of decision making. It must be clearly understood to separate the strategic and tactical decisions. Operations makes the tactical decisions of running the facility. Management makes the strategic decisions of assessing the facility's suitability for the job.

As we drive our cars, our driving habits and the outside conditions dictate how we act. If we get special causes and common causes confused, we could wind up trying to control the weather conditions. Since weather is outside our control and abilities, we would be trying to change the weather (an impossible task) and not trying to reduce our speed. The more we spend our energies trying to adjust the weather, the less time we have to actually drive. This would make our dangerous conditions even worse.

Please refer to Chapter 2 for a complete discussion of Division of Labor. There we rigorously discussed the potential violations to the second principle. Since tampering is generally caused by a lack of understanding of the Second Principle of Process Management (Division of Labor), we must assure that all appreciate this potential cause. Management must assure and take the responsibility that all associates understand Division of Labor.

Many questions about the chain of events that cause an effect must be researched. The real art of causal analysis and linking is determining which of these questions warrant an investment of the time and money required for further research. In many ways the research is like wildcatting for oil. Many causes will be confirmed. Many causes will be rejected. Often the research leads to more questions. The objective is to continue to build our Knowledge Base. This information can actually become a bigger asset than the manufacturing facility!

What is THE cause of an event? Some people rush to explain the cause of an event as one single thing. These people talk as if the explanation for the event can be determined by simply tracing a single line from a individual action. However, we are flirting with disaster by using overly simplistic reasoning. Instead, we must be vigilant for multiple causes and interactions in order to improve our Knowledge Base.

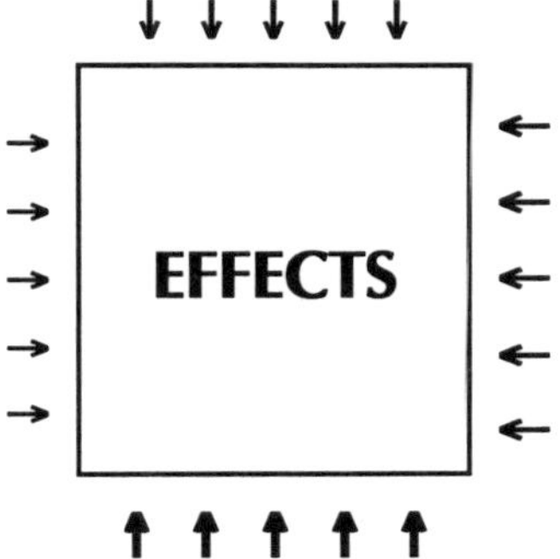

Assessing the Impact of the Change
Management as the strategic decision-maker must assess the impact of the change. While the tactical work force is working to determine the cause of the change, the strategist must assess the change through a capability study. The strategist's capability study will be covered in depth in Chapter 10.

Decisions for Action
The control chart is not a tool for action. The action must be driven from the determination of the cause and the results of the capability study.

Improvement
If the capability has improved and the cause found, then the cause is made a standard practice and the control chart limits are recalibrated. If the capability has improved and the cause cannot be found, then the best course of action is to bring more resources into play to continue and increase our search for the cause because a missed opportunity for improvement is here.

Deterioration
If the capability has deteriorated and the cause found, then the cause must be fixed and the process brought back to its original method. If the cause cannot be fixed, then the control chart limits must be recalibrated because this level of performance is now the norm for this machine.

No Cause Found
If no cause is found, our challenge is to continually search for the cause. This situation will occur many times, but perseverance is essential to this effort. In my boyhood television days, the Superman episodes always talked about a never ending search for truth, justice, and the American way. Our search for causes must be like Superman searching for the criminal: a never ending search for the cause of our change.

Remember, never take action just because an alarm is noted. The cause must always be found and the capability assessed before taking action.

Chapter Summary

Summary Of Control Chart Interpretation Guidelines

- ☞ Single point outside the control limits
- ☞ A series of points outside the control limits
- ☞ Seven consecutive points increasing (or decreasing) from the last point
- ☞ Seven consecutive points above (or below) the average
- ☞ More than 68% of the points in the first zone
- ☞ More than 32% of the points outside the first zone

CAUSES

SPECIAL causes are within tactical control.

- Tactical Action is required to deal with special causes.
- Can be resolved by people (tactician) closest to the process.

COMMON causes are strategic control.

Strategic Action is required to deal with common causes.

- Common to all operators and machines
- Require strategic action for resolution.

TAMPERING causes are caused by violations of Division of Labor.

- All associates must appreciate and understand their role and then execute the role.

Remember, never take action just because an alarm is noted. The cause must always be found and the capability assessed before taking action.

Chapter Eight: Other Variable Control Charts

We must now expand into other types of variable data. Chapter Seven covered the interpretation, reading, and use of control charts by covering the $\bar{X}$ and R charts in detail. Since all control charts are based on normal distribution, the remaining charts are used and read in the same way. Remember, the overall objective of control chart methods is to detect product or process change. The term **out of control** simply means a change has been detected.

The initial control limit calculations were covered in Chapter Six. From Chapter Six all the methods to effectively control a process and product were discussed. As a review, the following steps were covered:

- Calculate initial control limits
- Establish that the process is in control
- Continually survey the control charts for alarms (out-of-control conditions)
- Assess the capability of the in-control process

In Chapter Seven the use and reading of control charts was covered. The following summarizes the key points of Chapter Seven:

- **Establish that the process is in control** As each out-of-control condition is identified and dealt with, this procedure is repeated until no alarms have been detected. When the process is in control (no alarms have been detected), the control limits are locked in place. We must now explain the reading and interpretation of the control charts.

- **Continually survey the control charts for alarms (out-of-control conditions)** Data is gathered to monitor against the alarm mechanism (control limits) for changes in the conditions of the machine. When an out-of-control or change is detected the chart has done its job. The following tasks must be undertaken:
  i. Investigate for the cause
  ii. Alert management to the change
  iii. Decide what to do

Chapter Eight will now expand what we have already learned about SPC and control charts and add knowledge about the remaining classic variable control charts. Each set was designed to monitor for a particular situation. The remaining two classic charts are $\bar{X}$ & S and individual charts.

Since all of the classic control charts are comprised of two control charts, our knowledge of the $\bar{X}$ & R charts and its similarity to the others will allow us to expedite the explanation of how to construct the other charts. One chart monitors central tendency for any changes while the other control chart monitors for changes in the variability of the process. The table below shows what each chart is designed to monitor.

	Central Tendency	Variability	Distribution
$\bar{X}$ ***& R***	$\bar{X}$	***R***	***Central Limit Theorem***
	This chart was covered in depth in Chapter Six.		
$\bar{X}$ & S	$\bar{X}$	S	Central Limit Theorem
Individual	X	Moving Range	Histogram

The table above shows each chart monitoring its statistical element. The $\bar{X}$ and X control charts monitor central tendency and only the calculations are different. The R, S, and moving range all monitor variability for different situations using a specific calculation for the particular situation.

Remember, do not confuse control charts and control limits with a management chart to assess customer specifications. Control limits help us determine if a process is producing consistent output: that is, whether its distribution is stable and the process is in control and not changing.

Now we must focus on the mechanics of the remaining charts. The remaining charts then are continuations of this same approach. Let's now focus on the control chart called $\bar{X}$ & S.

$\bar{x}$ and S Charts

When To Use: These charts should be used when the number of samples in the subgroup is large and the **R** value is not precise. This math associated with standard deviation adds no more precision than the R value because the number of samples in the subgroup is small. Either control chart ($\bar{X}$ & R or $\bar{X}$ & S) will function well in most situations.

Since the R chart is simpler than the S chart, the S chart is very seldom used. The S chart will be the tool of choice when the number of samples in the subgroup is large. Large is typically greater than ten.

What does an $\bar{X}$ & S chart look like? Notice the $\bar{X}$ chart below looks very much like the original chart we discussed in Chapter Six. Only the calculations have been slightly changed.

$\bar{X}$ Chart is designed to monitor the movement or changes in the **central tendency** of the process. Throughout this book the term $\bar{X}$ will be used for average of the subgroup and $\bar{\bar{X}}$ for average of the averages.

The **$\bar{X}$ Chart** is comprised of the following plot lines:

MONITOR:

$\bar{\bar{X}}$ which is the benchmark of the **central tendency** of the process.

UCL_{AVE} which is the **upper control limit** that is three (3) standard errors above the mean.

LCL_{AVE} which is the **lower control limit** that is three (3) standard errors below the mean.

DATA:

$\bar{X}$ which is the average of each sub-group and becomes the data point to be monitored.

This chart is an example of an **$\bar{X}$ chart used with S chart.**

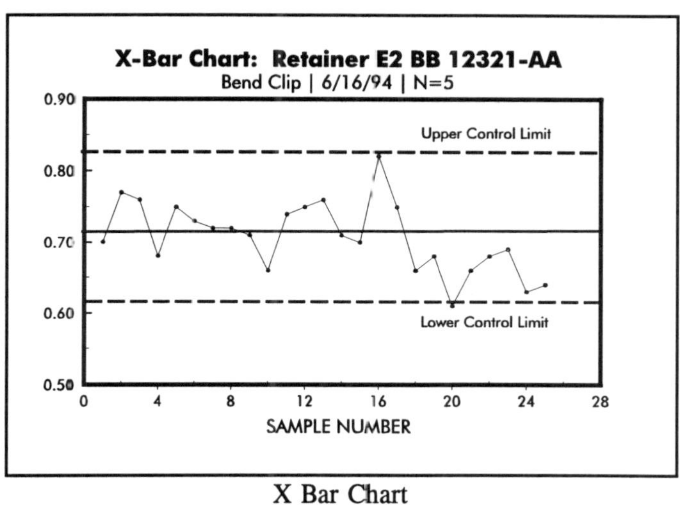

X Bar Chart

S Chart is designed to monitor the movement or changes in the variability of the process. This chart will track exactly like the R chart that has already been discussed. The S chart uses the standard deviation as the monitor while the R chart uses the R value. Only the math is different. The S chart is more precise if the number of samples in a subgroup is large. A large subgroup sample is very rare, thus making the S chart a rarity in actual practice.

The **S Chart** is comprised of the following plot lines:

MONITOR:

$\bar{S}$ which is the benchmark of the **variability** of the process.

UCL_s which is the **upper control limit** that is three (3) standard errors above the mean.

LCL_s which is the **lower control limit** that is three (3) standard errors below the mean.

DATA:

S which is the standard deviation of each sub-group and becomes the data point to be monitored.

This chart is an example of a S chart. Note how similar this chart is to the R chart that we discussed earlier.

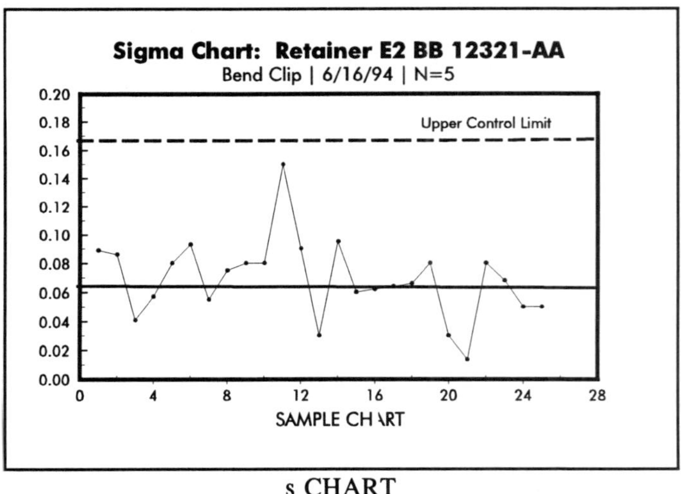

s CHART

$\bar{X}$ And S Chart Data The next table is the data that we will use for our $\bar{X}$ & S example. The data is from one of our candy plant machines from 1:00 to 4:00.

SUBGROUP	1:00	2:00	3:00	4:00	5:00	6:00
SAMPLE						
1	10	14	10	10		
2	11	8	10	11		
3	13	10	9	9		
4	9	10	10	10		
5	10	9	10	10		
6	8	13	11	11		
7	9	12	12	12		
8	9	11	8	8		
9	12	9	7	9		
10	10	9	10	10		

$\bar{X}$ And S Chart Data The following are the algebraic formulas that were used to compute subgroup control chart data points. They are used to derive the data to plot for the $\bar{X}$ and S charts. Average data is used to provide a means to alert operators when a shift in the process has occurred. This should eliminate the potential for over adjusting control or eliminating a test due to an overzealous alarm.

Remember, for control chart purposes the actual individual data is no longer of value and will not be used. These charts are tactical tools to assist the operator in the correct and consistent running of the machine.

$\overline{X}$ & S control chart data:

For our 1:00 through 4:00 data, the following formulas are used to compute data that will be charted. The following shows the data calculations for the 1:00 readings.

X_n is the data reading for the particular subgroup where n equals the size of the subgroup:

$$\overline{X} = \frac{X_1 + X_2 + X_3 + \ldots + X_N}{N}$$

where N equals the number of samples in the sub-group.

The actual 1:00 values are now inserted into the equation to compute the plotted 1:00 point for the control chart.

$$\overline{X} = \frac{10 + 11 + 13 + 9 + 10 + 8 + 9 + 9 + 12 + 10}{10}$$

$$= 10.1$$

This $\overline{X}$ value is now the 1:00 data point for plotting on the $\overline{X}$ chart. These calculations are then continued for all time frames. The actual individual data points are not used on the control charts.

Next the S value must be calculated for the 1:00 data points on the S chart.

$$s = \sqrt{\frac{\sum (\overline{X} - X_i)^2}{N - 1}}$$

The 1:00 data points are inserted into this equation to give the following.

$$s = \sqrt{\frac{(10.1 - 10)^2 + (10.1 - 11)^2 + \ldots + (10.1 - 10)^2}{10 - 1}}$$

The table on the next page shows the detail calculations for the 1:00 reading. The calculations are the distance away from the subgroup mean, the squared distance, sum of squares, the variance, and standard deviation.

Samples	X_{Ave}-X_i	Distance	Square of Distance
1	10.1 - 10	.1	.0
2	10.1 - 11	-.9	.8
3	10.1 - 13	-2.9	8.4
4	10.1 - 9	1.1	1.2
5	10.1 - 10	.1	.0
6	10.1 - 8	2.1	4.4
7	10.1 - 9	1.1	1.2
8	10.1 - 9	1.1	1.2
9	10.1 - 12	-1.9	3.6
10	10.1 - 10	.1	0.0

The last column of the table is the squared distance for each data point in the 1:00 subgroup. These ten values are summed to give the sum of the squares which is 20.9.

Next the number of samples for the 1:00 subgroup (ten) is subtracted by one to give a value of nine, called degrees of freedom. The sum of squares (20.9) is divided by the degrees of freedom (9) to calculate the variance of 2.3. The square root of the variance (2.3) is the standard deviation which will be referred to as S of 1.5. This S value is the 1:00 data point for the S chart and is seen below:

$$s = 1.5$$

This set of calculations is repeated for each subgroup and is shown in the table on the next page.

SUBGROUP	1:00	2:00	3:00	4:00	5:00	6:00
SAMPLE						
1	10	14	10	10		
2	11	8	10	11		
3	13	10	9	9		
4	9	10	10	10		
5	10	9	10	10		
6	8	13	11	11		
7	9	12	12	12		
8	9	11	8	8		
9	12	9	7	9		
10	10	9	10	10		
$\bar{X}$ = *Total/Samples*	10.1	10.5	9.7	10.0		
s	1.5	2.0	1.4	1.2		

On a control chart the actual individual candy readings will not be used. From this point on in this chapter, no use of the individual readings will be made. Therefore, in the tables that follow, all of the individual readings have been eliminated. When we return in Chapter Ten to discuss capability studies, the individual readings will become significant again.

$\bar{x}$ Chart Calculations for $\bar{x}$ and S

The following calculations are used to derive the control limits for the $\bar{X}$ Chart. The following pages show how our example uses these formulas.

X_n is the data reading for the particular sample where n equals the sample.

$$\bar{X} = \frac{X_1 + X_2 + X_3 + \ldots + X_N}{N}$$ where N equals the number of samples in the sub-group

$$\bar{\bar{X}} = \frac{\bar{X}_1 + \bar{X}_2 + \bar{X}_3 + \ldots + \bar{X}_K}{K}$$ where K equals the number of sub-groups

$$s = \sqrt{\frac{\sum (\bar{X} - X_i)^2}{N - 1}}$$

$$\bar{S} = \frac{s_1 + s_2 + s_3 + \ldots + s_K}{K}$$

$$UCL_{AVE} = \bar{\bar{X}} + A_3 \ \bar{S}$$ where (A_2) $(\bar{X})$ is an approximation of three (3) standard errors

$$LCL_{AVE} + \bar{\bar{X}} - A_3 \ \bar{S}$$ where (A_2) $(\bar{X})$ is an approximation of three (3) standard errors

A_3 is the conversion factor to arrive at three (3) standard errors for use in determining UCL & LCL of $\bar{X}$ chart.

c_4 is the conversion factor for determining standard deviation of the individuals. This will be used for capability studies and not for control limits.

Subgroup Size	A_3	C_4	B_3	B_4
10	0.975	0.9727	0.284	1.716
11	0.927	0.9754	0.321	1.679
12	0.886	0.9776	0.354	1.646
13	0.850	0.9794	0.382	1.618
14	0.817	0.9810	0.406	1.594
15	0.789	0.9823	0.428	1.572
16	.763	.9835	.448	1.552
17	.739	.9845	.466	1.534
18	.718	.9854	.482	1.518
19	.698	.9862	.497	1.503
20	.680	.9869	.510	1.490
21	.663	.9876	.523	1.477
22	.647	.9882	.534	1.466
23	.633	.9887	.545	1.455
24	.619	.9892	.555	1.445
25	.606	.9896	.565	1.435

XBar & S Conversion Tables

Control Limit Calculations

First, the average of the averages and the average S values are calculated. The following formulas are completed below.

$$\bar{X} = \frac{X_1 + X_2 + X_3 + \ldots + X_N}{N}$$ where N equals the number of samples in the sub-group

$$\bar{S} = \frac{s_1 + s_2 + s_3 + \ldots + s_K}{K}$$

$\bar{\bar{X}} = \frac{10.1 + 10.5 + 9.7 + 10.0}{4} = \frac{40.3}{4} = 10.08$
$\bar{S} = \frac{1.5 + 2.0 + 1.4 + 1.2}{4} = \frac{6.05}{4} = 1.51$

Next, the averages are used to compute the control limits by using the following equations. The constant A_3 is obtained from the table on the prior page. The subgroup size for our example is ten thus the A_3 constant that is used is 0.975.

$UCL_{AVE} = \bar{\bar{X}} + A_3 \; \bar{S}$ where (A_2) $(\bar{\mathbf{X}})$ is an approximation of three (3) standard errors

$LCL_{AVE} + \bar{\bar{X}} - A_3 \; \bar{S}$ where (A_2) $(\bar{\mathbf{X}})$ is an approximation of three (3) standard errors

$\bar{\bar{X}} = \frac{10.1 + 10.5 + 9.7 + 10.0}{4} = \frac{40.3}{4} = 10.08$
$\bar{S} = \frac{1.5 + 2.0 + 1.4 + 1.2}{4} = \frac{6.05}{4} = 1.51$
$UCL_{\bar{\bar{X}}} = \bar{\bar{X}} + A_3\bar{S} = 10.08 + .975(1.51) = 11.55$
$LCL_{\bar{\bar{X}}} = \bar{\bar{X}} - A_3\bar{S} = 10.08 - .975(1.51) = 8.60$

$\bar{x}$ Chart Example The table below shows all the details of the example. As an exercise, plot the values on the blank chart at the bottom of this page and compare your work to the completed chart later in the chapter.

Subgroup	1:00	2:00	3:00	4:00	5:00	6:00
Samples	10	10	10	10		
$\bar{X}$ = *Total / Samples*	10.1	10.5	9.7	10.0		
S	1.5	2.0	1.4	1.2		

$$\bar{\bar{X}} = \frac{10.1 + 10.5 + 9.7 + 10.0}{4} = \frac{40.3}{4} = 10.08$$

$$\bar{S} = \frac{1.5 + 2.0 + 1.4 + 1.2}{4} = \frac{6.05}{4} = 1.51$$

$$UCL_{\bar{\bar{X}}} = \bar{\bar{X}} + A_3\bar{S} = 10.08 + .975(1.51) = 11.55$$

$$LCL_{\bar{\bar{X}}} = \bar{\bar{X}} - A_3\bar{S} = 10.08 - .975(1.51) = 8.60$$

Complete the control chart on the graph paper below.

S Chart Example As you did earlier, plot the results on the graph below and compare your results to the graphs later in this chapter.

SUBGROUP	1:00	2:00	3:00	4:00	5:00	6:00
Samples	10	10	10	10		
S	1.5	2.0	1.4	1.2		

$$\overline{S} = \frac{1.5 + 2.0 + 1.4 + 1.2}{4} = \frac{6.05}{4} = 1.51$$

$$UCL_S = B_4\,\overline{S} = (1.716)(1.51) = 2.59$$

$$LCL_S = B_3\,\overline{S} = (0.284)(1.51) = .43$$

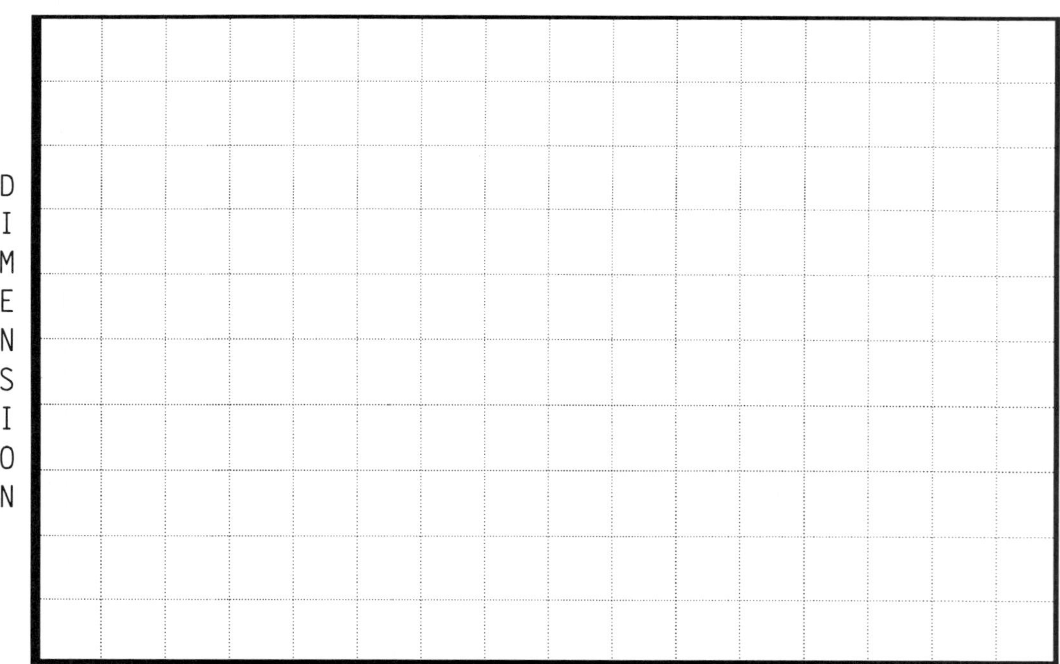

S Chart Calculations for $\bar{X}$ and s

The following calculations are used to derive the data plots for the **S Chart**:

$$s = \sqrt{\frac{\sum(\bar{X} - X_i)^2}{N - 1}}$$

$$\bar{S} = \frac{s_1 + s_2 + s_3 + \ldots + s_K}{K}$$

$UCL_s = B_4 \bar{S}$ where $(B_4)(\bar{S})$ is an approximation of three (3) standard errors above the mean.

$LCL_s = B_3 \bar{S}$ where $(B_3)(\bar{S})$ is an approximation of three (3) standard errors below the mean.

S Chart is designed to monitor the movement or changes in the variability of the process. The S Chart is comprised of the following plot lines:

$\bar{S}$ which is the benchmark of the center of the variability of the metric.
UCL_s which is the **upper control limit** above the benchmark variability.
LCL_s which is the **lower control limit** below the benchmark variability.

S_K is the variability monitor of the sub-group and becomes the data point to be plotted.

B_3 is the conversion factor to arrive at three (3) standard errors for use in determining LCL of S chart.

B_4 is the conversion factor to arrive at three (3) standard errors for use in determining UCL of S chart.

Subgroup Size	A_3	C_4	B_3	B_4
10	0.975	0.9727	0.284	1.716
11	0.927	0.9754	0.321	1.679
12	0.886	0.9776	0.354	1.646
13	0.850	0.9794	0.382	1.618
14	0.817	0.9810	0.406	1.594
15	0.789	0.9823	0.428	1.572
16	.763	.9835	.448	1.552
17	.739	.9845	.466	1.534
18	.718	.9854	.482	1.518
19	.698	.9862	.497	1.503
20	.680	.9869	.510	1.490
21	.663	.9876	.523	1.477
22	.647	.9882	.534	1.466
23	.633	.9887	.545	1.455
24	.619	.9892	.555	1.445
25	.606	.9896	.565	1.435

XBar and S Conversion Chart

$\bar{x}$ and S Charts

The $\overline{\mathbf{X}}$ **Chart** is designed to monitor the movement or changes in the **central tendency** of the process. Throughout this book the term $\overline{\mathbf{X}}$ will be used for average of the subgroup and $\overline{\overline{\mathbf{X}}}$ for average of the averages. The $\overline{\mathbf{X}}$ **Chart** is comprised of the following plot lines:

Monitor: $\overline{\overline{\mathbf{X}}}$ which is the benchmark of the **central tendency** of the process, UCL_{AVE} which is the **upper control limit** that is three (3) standard errors above the mean, LCL_{AVE} which is the **lower control limit** that is three (3) standard errors below the mean.

Data: $\overline{\mathbf{X}}$ which is the average of each sub-group and becomes the data point to be monitored.

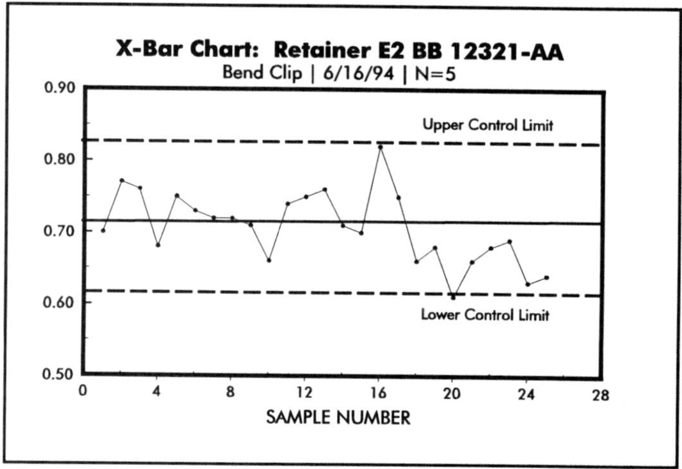

X Bar Chart

S Chart is designed to monitor the movement or changes in the variability of the process. The **S Chart** is comprised of the following plot lines:

Monitor: $\overline{\mathbf{S}}$ is the benchmark of variability, UCL_S is the **upper control limit** that is three (3) standard deviations above the mean, LCL_s which is the **lower control limit** that is three (3) standard errors below the mean.

Data: S which is the standard error of each sub-group and becomes the data point to be monitored.

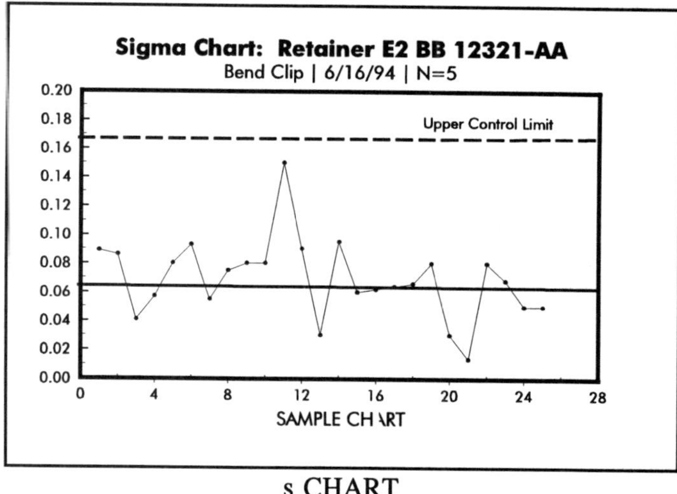

s CHART

Individual

When to Use: These charts should be used when we are able to take only one (1) sample. Examples of this situation would include chemical processes, expensive products, low volumes of product, or homogeneous readings.

Individual Chart is designed to monitor the movement or changes in the **central tendency** of the process. Throughout this book the term $\overline{\mathbf{X}}$ will be used for average.

The **X Chart** is comprised of the following plot lines:

MONITOR:

$\overline{\mathbf{X}}$ which is the benchmark of the **central tendency** of the process.

UCL_{AVE} which is the **upper control limit** that is three (3) standard errors above the mean.

LCL_{AVE} which is the **lower control limit** that is three (3) standard errors below the mean.

DATA:

X which is the data point to be monitored.

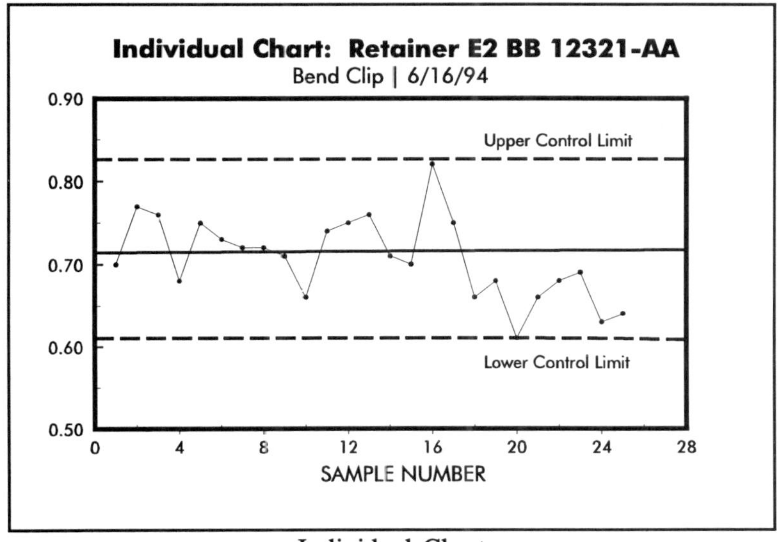

Individual Chart

Moving Range Chart is designed to monitor the movement or changes in the variability of the process.

The **Moving Range Chart** is comprised by the following plot lines:

MONITOR:

$\bar{\mathbf{R}}$ which is the benchmark of the **variability** of the process.

UCL_R which is the **upper control limit** that is three (3) standard errors above the mean.

LCL_R which is the **lower control limit** that is three (3) standard errors below the mean.

DATA:

R which is the moving range between the current and last reading and becomes the data point to be monitored.

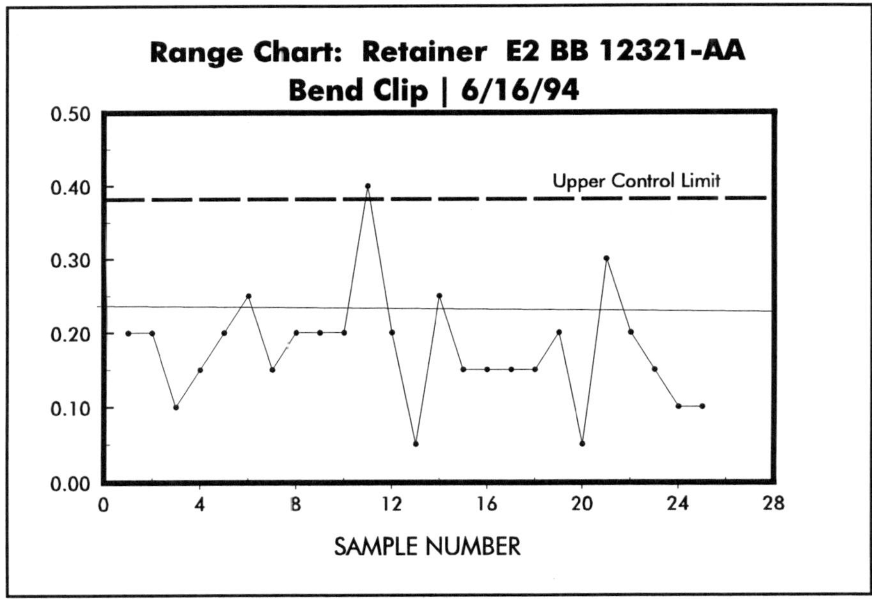

Moving Range Chart

A Histogram is used to monitor the distribution of the process in order to assure that it is truly normal.

X Chart Data

The data for the X chart of the individual chart will be the individual values. For the variability of the data a moving range will be used which compares the highest and lowest value to compute the distance between the values.

The table below shows the 1:00, 2:00, 3:00, 4:00, and 5:00 readings. These readings will be plotted on the individual chart as data.

SUBGROUP	1:00	2:00	3:00	4:00	5:00	6:00
SAMPLE						
1	10	12	14	15	13	

The moving range value is now computed for each time period for monitoring the variability of the values. The 1:00 reading has no prior reading so no moving range can be computed. Our first moving range is computed for the 2:00 reading, where the 2:00 data of 12 is compared to 1:00 reading of 10. The moving range at 2:00 is 2 which is plotted on the moving range chart. The formula for the 1:00 moving range is shown below:

$$R = | X_{Current} - X_{Prior} |$$

$$= | 10-12 |$$

$$= 2$$

Notice that the absolute value or positive value of the range is always used. This keeps the distance, whether moving up or down, relative. The table below shows the individual data and the moving range for all the example data.

SUBGROUP	1:00	2:00	3:00	4:00	5:00	6:00
SAMPLE						
1	10	12	14	15	13	
R		2	2	1	2	

The formula for the individual chart is shown on the next page.

X Chart Calculations - Individual

The following calculations are used to derive the data plots for the **X Chart**:

X_n is the data reading for the particular sample.

$$\bar{X} = \frac{X_1 + X_2 + X_3 \ldots + X_N}{N}$$ where N equals the number of samples

$$R = \left| X_{Current} - X_{Prior} \right|$$

$$\bar{R} = \frac{R_1 + R_2 + R_3 + \ldots + R_K}{K}$$ where K equals the number of sub-groups

Remember even though this is an individual chart, we are still using standard error for our computations of control limits.

$$UCL_{AVE} = \bar{X} + 2.660 \ \bar{R}$$

$$LCL_{AVE} = \bar{X} - 2.660 \ \bar{R}$$

X Chart Example

Now we must take our individual values and compute the individual control charts alarm system.

SUBGROUP	1:00	2:00	3:00	4:00	5:00	6:00
SAMPLE						
1	10	12	14	15	13	
R		2	2	1	2	

As we have done in all our other control charts, we first must compute the average of both the individuals and the average of the range using the following formula.

$$\bar{X} = \frac{X_1 + X_2 + X_3 \ldots + X_N}{N}$$ where N equals the number of samples

$$R = \left| X_{Current} - X_{Prior} \right|$$

$$\bar{R} = \frac{R_1 + R_2 + R_3 + \ldots + R_K}{K}$$ where K equals the number of sub-groups

$\bar{X} = \frac{10 + 12 + 14 + 15 + 13}{5} = \frac{64}{5} = 12.8$
$\bar{R} = \frac{2 + 2 + 1 + 2}{4} = \frac{7}{4} = 1.75$

With the averages computed, control limits can now be computed. The individual chart, like all other control charts, use standard error for computing the control limits. Since the moving average uses two values the sample size of the subgroup is two. To simplify the calculations the following factors and formula are used.

$$UCL_{AVE} = \bar{X} + 2.660\ \bar{R}$$
$$LCL_{AVE} = \bar{X} - 2.660\ \bar{R}$$

$\bar{X} = \frac{10 + 12 + 14 + 15 + 13}{5} = \frac{64}{5} = 12.8$
$\bar{R} = \frac{2 + 2 + 1 + 2}{4} = \frac{7}{4} = 1.75$
$UCL_{\bar{\bar{X}}} = \bar{X} + 2.660\ \bar{R} = 12.8 + (2.66)(1.75) = 17.46$
$LCL_{\bar{\bar{X}}} = \bar{X} - 2.660\ \bar{R} = 12.8 - (2.66)(1.75) = 8.15$

The individual control chart can now be plotted. In the space provided below complete the individual control chart using our example data.

SUBGROUP	1:00	2:00	3:00	4:00	5:00	6:00
SAMPLE						
1	10	12	14	15	13	
R		2	2	1	2	

$$\bar{X} = \frac{10 + 12 + 14 + 15 + 13}{5} = \frac{64}{5} = 12.8$$

$$UCL_{\bar{\bar{X}}} = \bar{X} + 2.660\ \bar{R} = 12.8 + (2.66)(1.75) = 17.46$$

$$LCL_{\bar{\bar{X}}} = \bar{X} - 2.660\ \bar{R} = 12.8 - (2.66)(1.75) = 8.15$$

$$\bar{R} = \frac{2 + 2 + 1 + 2}{4} = \frac{7}{4} = 1.75$$

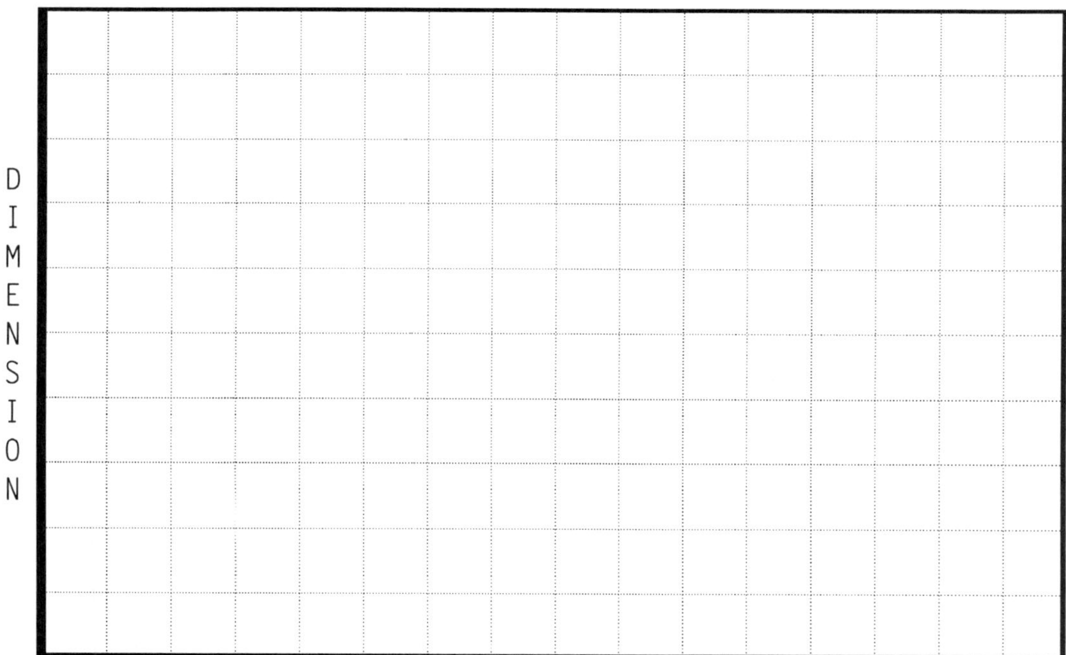

X Chart Calculations - Individual

In the previous example, the Moving Range was calculated based on the present and previous readings (a "Range Frequency" of *two* data points). Special cases call for a Moving Range to be figured based on three or even more data points. The formulas below are for calculating **individual chart alarms** for the generalized case of range frequency. If the moving range is based on a Range Frequency of more than two, then chart alarms should be calculated by using the following formulas and constants.

The following calculations are used to derive the data plots for the **individual chart**:

X_n is the data reading for the particular sample.

$$\bar{X} = \frac{X_1 + X_2 + X_3 \ldots + X_N}{N}$$ where N equals the number of samples

$$R = | X_i - X_j |$$

$$\bar{R} = \frac{R_1 + R_2 + R_3 + \ldots + R_K}{K}$$ where K equals the number of sub-groups

$$UCL_{AVE} = \bar{X} + (E_2)(\bar{R})$$ where $(E_2)(\bar{R})$ is an approximation of 3 standard errors

$$LCL_{AVE} = \bar{X} - (E_2)(\bar{R})$$ where $(E_2)(\bar{R})$ is an approximation of 3 standard errors

E_2 is the conversion factor to arrive at three (3) standard errors for use in determining UCL & LCL for Individual charts.

d_2 is the conversion factor for determining standard deviation of the individuals for use in a capability study. This constant is not used for the control charts.

Subgroup Freq	E_2	d_2	D_3	D_4
2	2.660	1.128	-	3.267
3	1.772	1.693	-	2.574
4	1.457	2.059	-	2.282
5	1.290	2.326	-	2.114
6	1.184	2.534	-	2.004
7	1.109	2.704	0.076	1.924
8	1.054	2.847	0.136	1.864
9	1.010	2.970	0.184	1.816
10	0.975	3.078	0.223	1.777

Individual Chart Conversion

Moving Range Data

The control chart limits can now be calculated.

SUBGROUP	1:00	2:00	3:00	4:00	5:00	6:00
SAMPLE						
1	10	12	14	15	13	
R		2	2	1	2	

The following are formulas for computing the alarm system for the moving range control chart:

$UCL_R = (D_4)(\bar{R})$ where (D_4) $(\bar{R})$ is an approximation of three (3) standard errors

$$UCL_R = (D_4)(\bar{R})$$
$$=(3.267)(1.75)$$
$$= 5.72$$

$LCL_R = (D_3)$ $(\bar{R})$ where (D_3) $(\bar{R})$ is an approximation of three (3) standard errors

$$LCL_R = (D_3)(\bar{R})$$
$$= 0$$

The table below shows the results of our calculations.

$\bar{R} = \frac{2 + 2 + 1 + 2}{4} = \frac{7}{4} = 1.75$
$UCL_R = (3.267)\ \bar{R} = (3.267)(1.75) = 5.72$
$LCL_R = 0$

The next page shows the general formula for a moving range control chart.

Moving R Chart Calculations - Individual

The following calculations are used to derive the data plots for the Moving Range Chart. Since most individual charts are based on the current and prior values, then:

$$R = | X_{Current} - X_{Prior} |$$

$$\bar{R} = \frac{R_1 + R_2 + R_3 + \ldots R_K}{K}$$

$$UCL_R = 3.267 \; \bar{R}$$ where K equals the number of data points

$$LCL_R = 0$$

R Chart is designed to monitor the movement or changes in the variability of the process. The **R Chart** is comprised of the following plot lines:

- $\bar{\mathbf{R}}$ which is the benchmark of the variability
- UCL_R which is the **upper control limit** above the benchmark variability.
- LCL_R which is the **lower control limit** below the benchmark variability.
- R_K which is the variability of the sub-group.

Moving Range Example

The moving range control chart can now be plotted. In the space provided below, complete the individual control chart using our example data.

SUBGROUP	1:00	2:00	3:00	4:00	5:00	6:00
SAMPLE						
1	10	12	14	15	13	
R		2	2	1	2	

$$\bar{R} = \frac{2 + 2 + 1 + 2}{4} = \frac{7}{4} = 1.75$$

$$UCL_R = (3.267)\ \bar{R} = (3.267)(1.75) = 5.72$$

$$LCL_R = 0$$

Use the space below to chart the moving range control chart for our example.

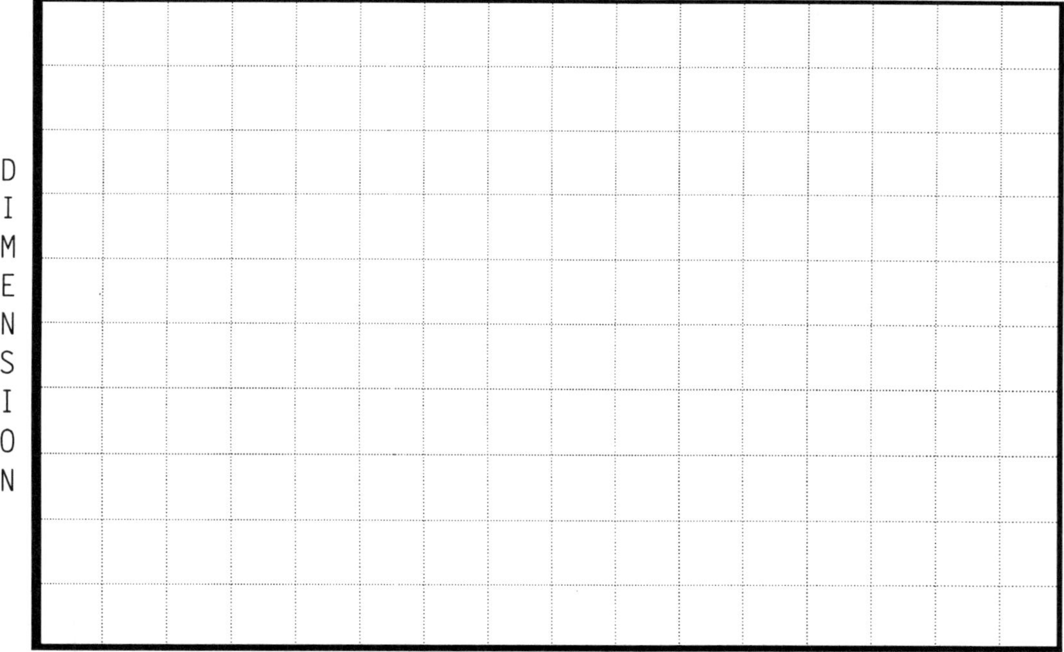

Moving R Chart Calculations - Individual

The following are the generalized calculations for varying range frequency. If the moving range is based on a Range Frequency of more than two, then chart alarms should be calculated by using the following formulas and constants. The following calculations are used to derive the data plots for the **R Chart**:

$$R = \left| X_{Current} - X_{Prior} \right|$$

$$\bar{R} = \frac{R_1 + R_2 + R_3 + \ldots R_K}{K}$$ where K equals the number of the data points

$$UCL_R = (D_4)(\bar{R})$$ where (D_4) $(\bar{R})$ is an approximation of three (3) standard errors

$LCL_R = (D_3)\ (\bar{R})$ where (D_3) $(\bar{R})$ is an approximation of three (3) standard errors

R Chart is designed to monitor the movement or changes in the variability of the process. The R Chart is comprised of the following plot lines:

$\bar{R}$ which is the benchmark of the variability

UCL_R which is the **upper control limit** above the benchmark variability.

LCL_R which is the **lower control limit** below the benchmark variability.

R_K which is the variability of the sub-group.

CAUTION! Once the process is in control stop recalculating the control limits.

D_3 is the conversion factor to arrive at three (3) standard errors for use in determining LCL of R chart.

D_4 is the conversion factor to arrive at three (3) standard errors for use in determining UCL of R chart.

Subgroup Freq	E_2	d_2	D_3	D_4
2	2.660	1.128	-	3.267
3	1.772	1.693	-	2.574
4	1.457	2.059	-	2.282
5	1.290	2.326	-	2.114
6	1.184	2.534	-	2.004
7	1.109	2.704	0.076	1.924
8	1.054	2.847	0.136	1.864
9	1.010	2.970	0.184	1.816
10	0.975	3.078	0.223	1.777

Individual Chart Conversion

Individual

Individual Chart is designed to monitor the movement or changes in the **Central Tendency** of the process. Throughout this book the term $\overline{\mathbf{X}}$ will be used for average. The individual chart is comprised of the following plot lines:

Monitor: $\overline{\mathbf{X}}$ which is the benchmark of the **central tendency** of the process, UCL_{AVE} which is the **upper control limit** that is three (3) standard errors above the mean, LCL_{AVE} which is the **lower control limit** that is three (3) standard errors below the mean.

Data: X which is the data point to be monitored

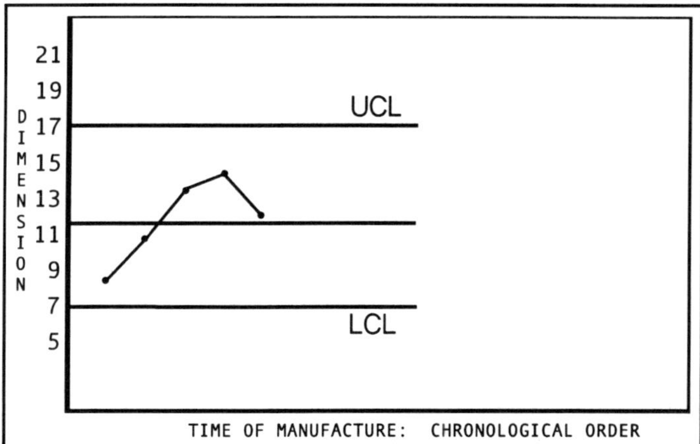

XBAR CHART
(Individual Chart)

Moving Range Chart is designed to monitor the movement or changes in the variability of the process. The Moving Range Chart is comprised of the following plot lines:

Monitor: $\overline{\mathbf{R}}$ which is the benchmark of the **central tendency** of the process, UCL_R which is the **upper control limit** that is three (3) standard errors above the mean, LCL_R which is the **lower control limit** that is three (3) standard errors below the mean.

Data: R which is the moving range between the current and last reading and becomes the data point to be monitored.

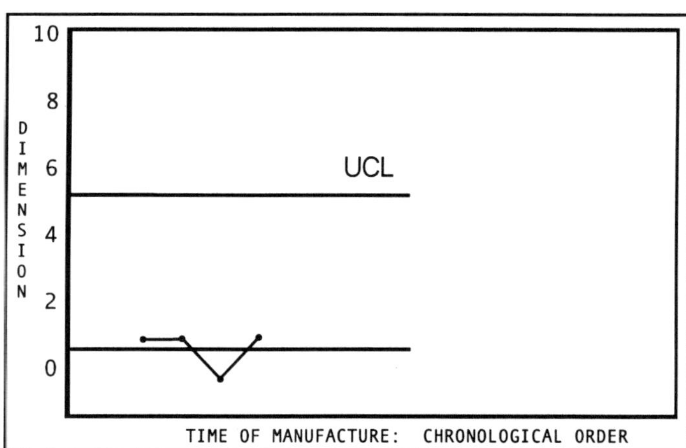

Moving Range Chart

Now that we have covered the classic variable charts, let's recap where the different charts are to be used. Each chart was designed for a particular situation, and all detect a shift in the metric that they are monitoring.

The $\bar{X}$ & R chart is designed for a high volume operation and is the most commonly used chart. The $\bar{X}$ & S chart takes care of the case where large samples are pulled for each subgroup. The individual charts monitor situations where only one point is pulled.

Let's cover the some special kinds of metrics and which control chart to use.

When to use which charts

The chart below summarizes our variable control charts and the situations that require their use.

	$\bar{X}$ & R	$\bar{X}$ & S	Individual
Product Metrics: High volume. Discrete product.	X		
Product Metrics: High volume. Continuous product.			X
Product Metrics: High volume. Discrete product. Large samples in a subgroup.		X	
Product Metrics: Low volume			X
Product Metrics: Expensive product with destructive test.			X
Process Metrics: Continuous			X
Process Metrics: Discrete	X		

This chart provides some guidelines for the selection of the proper control chart to use for a particular situations. This should provide you with a starting point for the monitoring effort.

Chapter Summary

$\bar{X}$ and S Charts

$$\bar{S} = \frac{S_1 + S_2 + S_3 + \ldots + S_K}{K}$$

A_3 is the conversion factor to arrive at three (3) standard errors for use in determining UCL & LCL of $\bar{X}$ chart where $(A_2)(\bar{\mathbf{X}})$ is an approximation of three (3) standard errors

B_3 is the conversion factor for the approximation of three (3) standard errors used as the lower control limit on the s chart.

B_4 is the conversion factor for the approximation of three (3) standard errors used as the upper control limit on the s chart.

Subgroup Size	A_3	C_4	B_3	B_4
10	0.975	0.9727	0.284	1.716
11	0.927	0.9754	0.321	1.679
12	0.886	0.9776	0.354	1.646
13	0.850	0.9794	0.382	1.618
14	0.817	0.9810	0.406	1.594
15	0.789	0.9823	0.428	1.572
16	.763	.9835	.448	1.552
17	.739	.9845	.466	1.534
18	.718	.9854	.482	1.518
19	.698	.9862	.497	1.503
20	.680	.9869	.510	1.490
21	.663	.9876	.523	1.477
22	.647	.9882	.534	1.466
23	.633	.9887	.545	1.455
24	.619	.9892	.555	1.445
25	.606	.9896	.565	1.435

$\bar{X}$ Chart Control Limits

$$UCL_{AVE} = \bar{\bar{X}} + A_3 \bar{S}$$

$$LCL_{AVE} + \bar{\bar{X}} - A_3 \bar{S}$$

S Chart Control Limits

$$UCL_s = B_4 \bar{S}$$

$$LCL_s = B_3 \bar{S}$$

Individual Charts

$$R = | X_i - X_j |$$

$$\bar{R} = \frac{R_1 + R_2 + R_3 + \ldots + R_K}{K}$$ where K equals the number of sub-groups

E_2 is the conversion factor to arrive at three (3) standard errors for use in determining UCL & LCL of Individual chart.

d_2 is the conversion factor for determining standard errors of the individuals.

D_3 is the conversion factor to arrive at three (3) standard errors for use in determining LCL of R chart.

Subgroup Freq	E_2	d_2	D_3	D_4
2	2.660	1.128	-	3.267
3	1.772	1.693	-	2.574
4	1.457	2.059	-	2.282
5	1.290	2.326	-	2.114
6	1.184	2.534	-	2.004
7	1.109	2.704	0.076	1.924
8	1.054	2.847	0.136	1.864
9	1.010	2.970	0.184	1.816
10	0.975	3.078	0.223	1.777

Individual Chart Conversion

D_4 is the conversion factor to arrive at three (3) standard errors for use in determining UCL of R chart.

$\bar{X}$ Chart Control Limits

$$UCL_{AVE} = \bar{X} + (E_2)(\bar{R})$$

$$LCL_{AVE} = \bar{X} - (E_2)(\bar{R})$$

Moving Range Control Limits

$$UCL_R = (D_4)(\bar{R})$$

$$LCL_R = (D_3)(\bar{R})$$

Remember all the variable charts are based on a normal distribution and use the same interpretation rules. We have covered the construction of the variable control charts and are now ready to move to Chapter Nine to show you how to construct attribute charts.

Chapter Nine: Attribute Control Charts

In Chapter Five *"Where Do We Start?"*, the two types of metric measurements of variable and attribute data were covered. In Chapters Six and Eight, the variable measurement control charts were constructed and covered in depth. Now we must move on to attribute control charts.

Remember, a control chart is the operator's tool to monitor the workings of a particular machine or system in chronological order. Attribute control charts are often confused with management reports, but like all other control charts, they are still a tactical tool. Attribute control charts include an alarm system with a mechanism for detecting any type of shift or change. This alarm must be dynamically re-calibrated to reflect the current status.

The attribute metric measurement type of data is generally thought of as the acceptance (or rejection) of a part. Even though this type of data is pass/fail, we can count the frequency of occurrence. Attribute (categorical) data is "yes" or "no", "good" or "bad", "go" or "no go".

What are the Attribute Control Charts?

NP Charts measure the frequency of the non-conforming parts. The inspection sample sizes (N) for this chart must be equal.

P Charts measure the proportion of units non-conforming.

C Charts measure the frequency or number of non-conformities (defects) where C equals the number of defects.

U Charts measure the proportion of defects by counting the number of non-conformities (defects) (C) and dividing by the sample size (N).

The attribute control charts' objective is to detect product or process change. The demonstrated effectiveness of a control chart is attributable to its objective and systematic analytical methods. The methods for effective control of attribute data are the same as all the other control charts and include the following steps:

- Calculate initial control limits
- Establish that the process is in control
- Continually survey the control charts for alarms (out-of-control conditions)
- Assess the capability of the in-control process

Calculate initial control limits To set the initial control limits, data must be gathered. As a guideline, thirty sets of values are captured to calculate and post on the charts. Remember, since a control chart is for one particular machine, the data must come from that one machine. Different control alarms for each machine would not be surprising. These same thirty subgroups are then used to compute trial control limits. The subgroup reading and the alarms are then posted on the chart for statistical control interpretation. This chapter will explain the mechanics and calculations for control limits.

Chapter Seven covered the interpretation, reading, and the use of control charts. These details of control chart interpretation tasks are used now for attribute data. From Chapter Seven's information, the following was explained and can be used for all our attribute charts because the attribute charts use a special type of normal distribution.

- **Establish that the process is in control** As each out of control condition is identified and dealt with, this procedure is repeated until no alarms have been detected. When the process is in control (no alarms have been detected), the control limits are locked in place. Chapter Seven explained the reading and interpretation of the control charts.

- **Continually survey control charts for alarms (out-of-control conditions)** Data is gathered to monitor against the alarm mechanism (control limits) for changes in the conditions of the machine. When an out-of-control situation or change is detected, the chart has done its job. The following tasks then must be undertaken:
  i. Investigate the cause.
  ii. Alert management to the change.
  iii. Decide what to do.

What is Attribute Data?

Many situations require attribute control charts. These situations are still for a continuous operation like all the other control charts. These tactical tools all must be used to monitor a continuous operation.

Attribute data may be gathered from many sources such as textile rolls of fabrics, the communications and information industries, the maintenance of a manufacturing plant, environmental information, transportation and traffic information, safety information, maintenance work orders for manufacturing and service industries, and the scrap rate for a manufacturing machine.

I, like most other business professionals today, must travel extensively. Almost all the stories I tell about my travels would make a good comedy, but most of my best stories have occurred at the airport or on a commercial airliner. Over the years I learned to accept the inconvenience of commercial air travel. Maybe rather than accept this, I should encourage the airlines to use statistical process control (SPC) to improve their processes.

The baggage handling and claim process as depicted to the right will be the example that we will expand into our exercises for explaining attribute control charts. If you have flown on a commercial airline, you already know that your baggage does not always arrive at your destination with you.

Now let's focus our attention on the details dealing with the construction needs of attribute charts. Our first attribute control chart is the NP chart. We will use the airline example to explain the attribute charts.

The baggage handling problem of monitoring the consistency of the lost bags will be our first problem. Each day on the 3:00 pm flight from Nashville, Tennessee to Orlando, Florida, 200 bags are processed. The following table shows the number of lost bags each day. On Monday two bags were lost, on Tuesday three bags were lost, on Wednesday four bags were mishandled, on Thursday there were three misplaced bags, and on Friday another three bags were lost.

SUBGROUP	Mon	Tues	Wed	Thurs	Fri	Sat
NP	2	3	4	3	3	

Our reason for monitoring the bags with a control chart is to determine how consistent the airlines baggage handling system really is. This control chart, like all the others, is a operational tool for the consistency of the baggage lost and to detect any changes in the consistency of the baggage handling system.

Since the total number of bags handled each day is the same (200 bags each day), then the NP control chart can be used as our monitoring system.

What is an NP Chart? A NP chart is an operator's tool to monitor the consistency of counts where the size in each group is the same. Remember, this chart like all other control charts, is intended to detect change for investigation. NP charts will have the same alarm system of calculated values that make up all control charts. This alarm system is calculated from past history to provide the center point and the control limits. The chart below is an example of a NP chart.

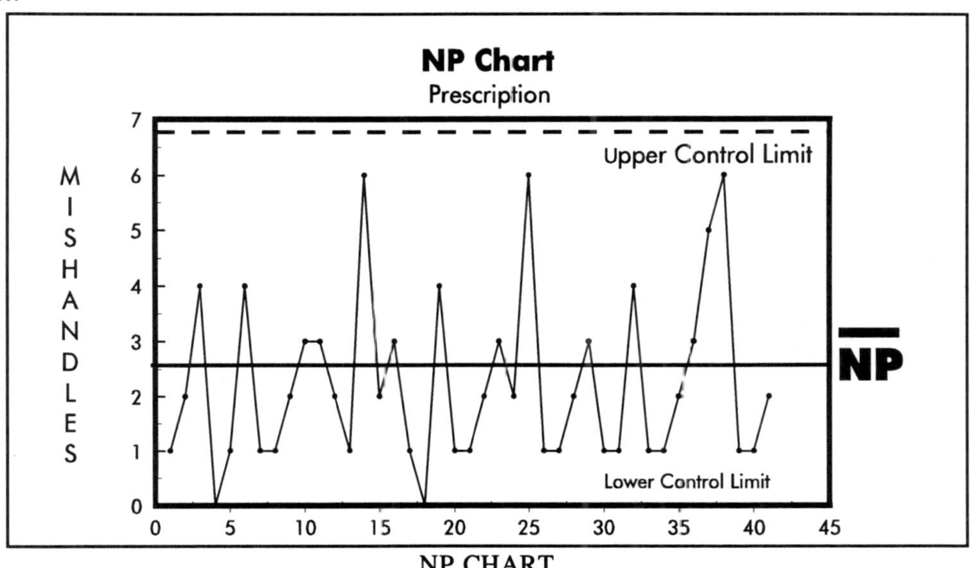

NP CHART

This page shows the formula for computing the control limits for NP charts.

NP Charts

When do we use the NP chart? NP Charts measure the frequency of the non-conforming parts. The inspection sample size (N) for this chart must be equal for each reading.

Data
NP equals the number of non-conforming parts.

N equals the subgroup sample size, and for an NP chart, the N value (size of the subgroup) must always be the same from group to group.

Alarms
The following formulas are used to compute the alarm system for both the center point and the control limits. The control limits for the NP chart are used and read like all the other control charts where only .1% of the value should fall above or .1% should fall below.

$$\bar{NP} = \frac{NP_1 + NP_2 + NP_3 + \ldots NP_K}{K}$$

where K equals the number of subgroups.

The upper and lower control limit formulas are shown below.

$$UCL_{NP} = \bar{NP} + 3\sqrt{\bar{NP}\left(1 - \frac{\bar{NP}}{N}\right)}$$

$$LCL_{NP} = \bar{NP} - 3\sqrt{\bar{NP}\left(1 - \frac{\bar{NP}}{N}\right)}$$

Remember, The NP chart can be used only when the sample size is equal for all subgroups. Also (like all other control charts) when the process or product is stable, the alarm system is locked in place until a decision is made to recalibrate the alarms that were identified from a shift in the process.

Let's take our airline baggage problem and see how the completed NP chart looks.

NP exercise

The NP values equal the number of non-conforming parts or lost bags for each day.

SUBGROUP	Mon	Tues	Wed	Thurs	Fri	Sat
NP	2	3	4	3	3	

The monitoring system for the NP chart for the 3:00 PM flight is shown below. N equals the subgroup sample size and must be present for the NP chart to perform correctly. For the airline example, the total number of bags or NP is 200 bags.

$$\bar{NP} = \frac{NP_1 + NP_2 + NP_3 + \ldots NP_K}{K}$$

$$= \frac{2 + 3 + 4 + 3 + 3}{5}$$

$$= \frac{15}{5}$$

$$= 3$$

where K equals the number of subgroups

The formula for computing the upper control limits is shown below.

$$UCL_{NP} = \bar{NP} + 3\sqrt{\bar{NP}\left(1 - \frac{\bar{NP}}{N}\right)}$$

Now the upper control limit calculation and results for the exercise above is shown below.

$$UCL_{NP} = 3 + 3\sqrt{3\left(1 - \frac{4}{200}\right)}$$

$$= 3 + 3\sqrt{3\ (1 - .02)}$$

$$= 3 + 3\sqrt{3(.98)}$$

Continuing to solve the equation we see the following results

$$UCL_{NP} = 3 + 3\ (1.98)$$

$$= 3 + 5.94$$

$$= 8.94$$

Finally, the upper control limit for our exercise is 8.94. Let's move onto the calculations for lower control limit.

The lower control limit calculation and results for the exercise above is shown below. For the lower control limit we have shortened the steps to the result.

$$LCL_{NP} = \bar{NP} - 3\sqrt{\bar{NP}\left(1 - \frac{\bar{NP}}{N}\right)}$$

$$= 3 - 3\sqrt{3\left(1 - \frac{4}{200}\right)}$$

$$= 3 - 5.94$$

$$= 0$$

The NP average, upper control limit, and lower control limit form an alarm system for detecting change and shifts.

Our exercise is now complete and the results are shown in the table below. Remember the NP chart is used as a monitor for detecting a change and shift for our airline bag handling exercise problem. This chart is not intended to assess whether the change is good or bad but simply that a change has occurred.

SUBGROUP	Mon	Tues	Wed	Thurs	Fri	Sat
NP	2	3	4	3	3	

$\overline{NP} = 3$
UCL = 8.94
LCL = 0

Now the completed NP chart for our example is shown below.

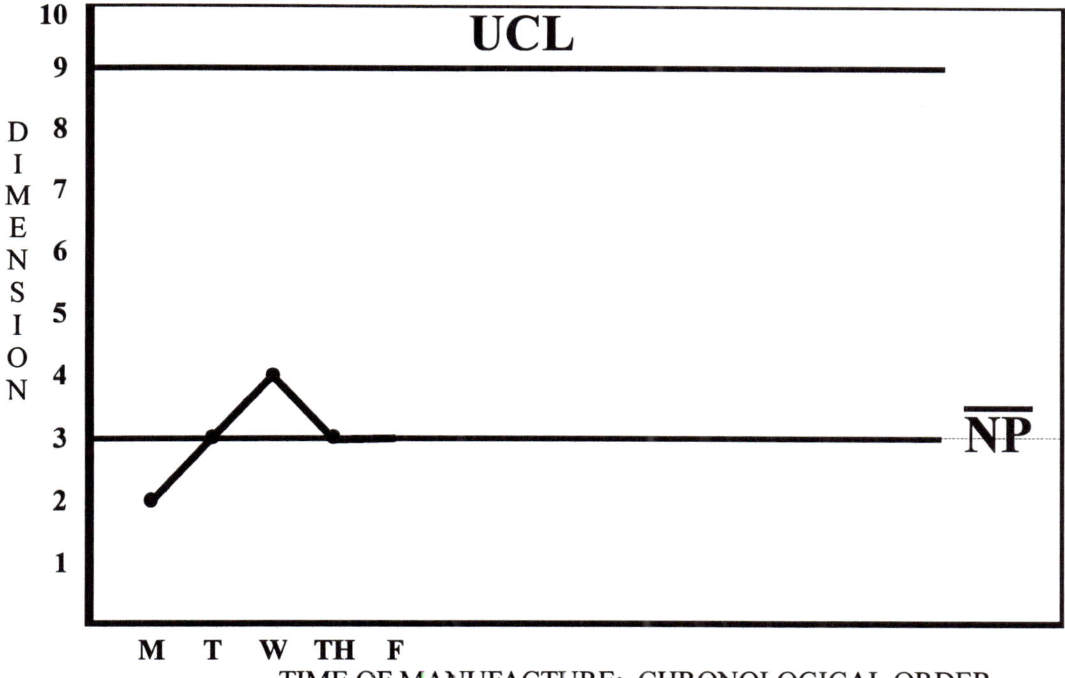

The interpretation rules for the NP chart are the same rules that were covered in Chapter Seven. For a detailed explanation of the rules see Chapter Seven. The explanation for why these interpretation rules work will be covered later in this chapter.

What if the Sample Size Varies?

The NP chart can only be used when the sample size is equal for each subgroup. Therefore, the NP chart is a wonderful tool, but it is limited by the requirement for subgroup size equality.

Airline Industry

For most airline flights from Nashville to Orlando, the total number of bags handled usually will not be the same from day to day. A more realistic problem is shown below where the total number of bags varies by day.

Two hundred bags are handled on Monday. 200 on Tuesday, 200 on Wednesday, 150 on Thursday, and 150 bags on Friday.

The total number of bags per daily flight is shown below.

SUBGROUP	Mon	Tues	Wed	Thurs	Fri	Sat
N	200	200	200	150	150	

To resolve the unequal number of samples in the subgroup a new control chart is required. This new control chart is called the **P chart** and is designed to resolve the unequal sample size issue.

How Does a P Chart Look?
The P chart uses the proportion of the count divided by the number of samples as a means of computing the data to plot. This ratio gives equal weight when the sample size changes. The chart below is an example of a P chart. This chart again, like all the control charts, is a tactical tool for detecting change.

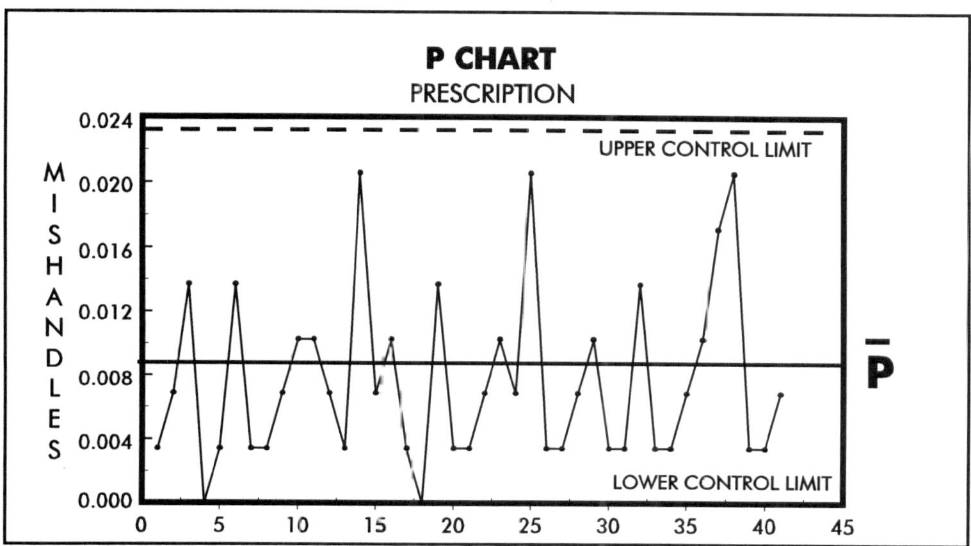

The P chart uses the P values (the ratios, proportions, or percentages) as the data. The P value resolves the unequal subgroup size problem. This historical data is then used to compute the alarm system of the center point and the control limits.

The control limits are the points outside of which we would be surprised to see a data point (P) fall. The P control limits use the same mathematical principles that all the other control charts use.

The P chart calculations and formulas are explained on the next page.

P Charts

P Charts measure the proportion of units non-conforming. By using proportion or ratios, rather than the NP counts, the unequal samples for the subgroups will be resolved.

DATA The data for the P chart is calculated by computing the ratio of the number of nonconforming items divided by the number of samples in the subgroup. This formula is shown below:

$$P = \frac{NP}{N}$$

where N equals number of items checked ; NP equals the number of items failing

MONITOR

The value and formula below is the center point at which, when the process is run correctly and consistently, the P values should be half on one side and half on the other.

$$\bar{P} = \frac{NP_1 + NP_2 + NP_3 \ldots NP_K}{N_1 + N_2 + N_3 \ldots N_K}$$

$$\bar{N} = \frac{N_1 + N_2 + N_3 + \ldots N_K}{K}$$

where K is the number of subgroups studied

Upper and Lower Control Limit

The upper and lower control limit formulas below are the points where a value falling outside is a rarity.

$$UCL_P = \bar{P} + 3\,\frac{\sqrt{\bar{P}\,(1-\bar{P})}}{\sqrt{\bar{N}}}$$

$$LCL_P = \bar{P} - 3\,\frac{\sqrt{\bar{P}\,(1-\bar{P})}}{\sqrt{\bar{N}}}$$

Don't forget, the P chart is used to detect change. When the process or product is stable, the alarm system is locked in place until a decision is made to recalibrate the alarms based on a change in the process.

P Chart Data

The NP chart works when the sample size is equal from subgroup to subgroup, but the total number of bags handled in the example below varies from day to day. This varying size is really a more typical life situation. Because of the typical shifting of the size, the P chart is used much more often than the NP chart is.

To have a relative number that is comparable from day to day then, the number of bags lost must be made relative to the total number of bags processed. Notice that three bags were lost on both Tuesday and Thursday. However, 200 bags were processed on Tuesday while only 150 bags were processed on Thursday. Therefore, a simple comparison of the lost bags will give an incorrect picture.

SUBGROUP	Mon	Tues	Wed	Thurs	Fri	Sat
NP	2	3	4	3	3	
N	200	200	200	150	150	

By making the lost bags a function of the total bags (NP/N), the relationship between the days is much clearer. Tuesday' ratio or P value is .015 and Thursday's ratio or P value is .02. The table below shows the computed results of the P values for each day.

SUBGROUP	Mon	Tues	Wed	Thurs	Fri	Sat
NP	2	3	4	3	3	
N	200	200	200	150	150	
P calculations	$\frac{2}{200}$	$\frac{3}{200}$	$\frac{4}{200}$	$\frac{3}{150}$	$\frac{3}{150}$	
P	.01	.015	.02	.02	.02	

Now the calculation and establishment of the control limits are required.

P Chart Example Control Limits

SUBGROUP	Mon	Tues	Wed	Thurs	Fri	Sat
NP	2	3	4	3	3	
N	200	200	200	150	150	
P	.01	.015	.02	.02	.02	

$$\bar{P} = \frac{2 + 3 + 4 + 3 + 3}{200 + 200 + 200 + 150 + 150} = \frac{15}{900} = .017$$

$$UCL = .017 + 3\frac{\sqrt{(.017)(1 - .017)}}{\sqrt{900 / 5}} = .017 + \frac{3(.1293)}{13.416} = .017 + .029 = .046$$

$$LCL = .017 - 3\frac{\sqrt{(.017)(1 - .017)}}{\sqrt{900 / 5}} = .017 - .029 = 0$$

Now the completed P chart for our example is shown below.

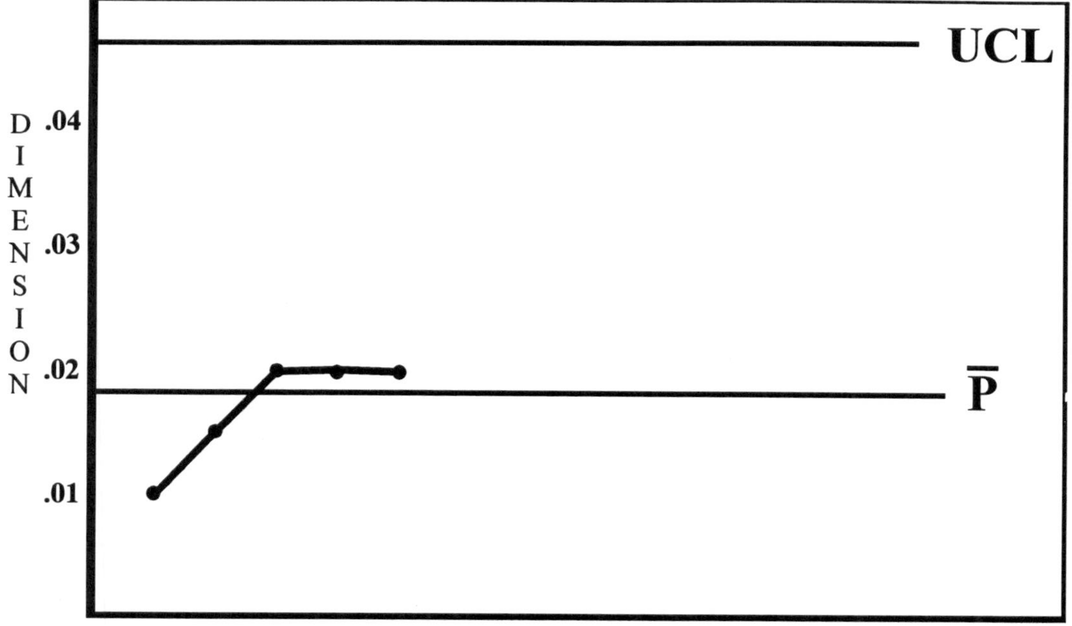

More About Attributes

Revisiting our good friend the normal distribution gives us a good starting point for discussing the distribution for attributes. The following is the earlier description of the normal or bell-shaped distribution.

Description Of The Normal Distribution

Average	...is the central point of the measurements.
Standard Deviation	...is the variability of the measurements.

The distribution is symmetrical around its average. Fifty percent of the data points are above the average and fifty percent of the measurements are below the average.

The frequency of occurrence drops rapidly as we get farther and farther away from the center. This characteristic is for both above and below the average.

The frequency of occurrence above and below the average are symmetrical.

As we move farther and farther away from the average, the number of occurrences drops at a rapid rate. This trait for the normal is in both directions from the average.

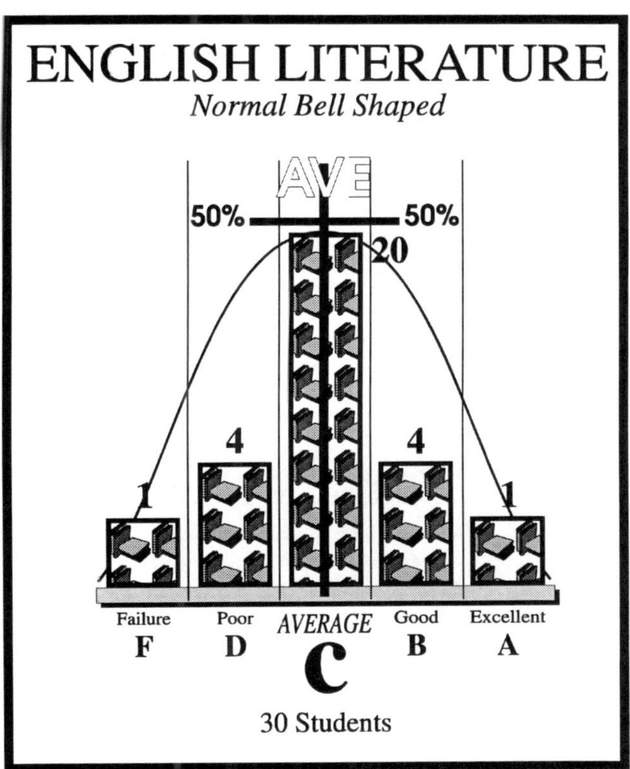

The chart to the right is used to depict how a process, with its measurements, will look. The chart shows the spread of grades from the English Literature class of thirty students covered in Chapter One. When the score measurements are converted to letter grades, the histogram of the process is well represented by a normal distribution. The perfect normal does not have the steps of a histogram but is a smooth curve A very clear pattern is formed for the probabilities of the normal distribution. These probabilities were used for the alarms associated with variable data.

Why is the Normal Distribution Important? Having data that is well represented by a normal distribution allows us to predict using the probabilities of the normal. A good understanding of these probabilities was an essential part of the alarms for all the variable control charts. How does the knowledge of the normal help us with the alarms for attribute data?

Attribute charts are based on a special kind of distribution called a **binomial**. The binomial is really a special kind of normal distribution. Let's discuss this binomial distribution.

Binomial Distribution: We need to develop a methodology or distribution to handle attribute data. The type of data we will analyze is an either/or condition like a true/false condition. The airline example of lost bags fits this category. The bag arrives with you and your flight (true) or the bag does not arrive with you and your flight (false). We will come back to our airline problem but first let's take a simpler example to clarify the binomial.

The example we will use is the tossing of a quarter and determining the value of heads or tails. The probability of heads occurring is 1/2. The flipping of the coin is a perfect example of the binomial. Head is the true and tails is the false side of the equation.

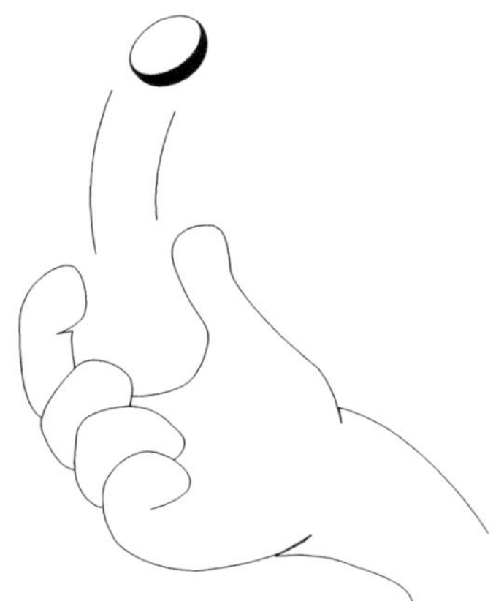

The binomial distribution has the following conditions:

1. A fixed number, n, of objects is examined.
2. The probability of occurrence is the same for each test.
3. Each test is independent.

For a binomial where n = the number of items and p = probability of the occurrence, the mean = p_{Ave}, and the *Standard Deviation* = $\sqrt{n\bar{p}(1-\bar{p})}$.

We will flip our quarter twenty times fulfilling the condition of a fixed number of trials.

Thus, our coin toss meets the three criteria of the binomial. We will flip the quarter twenty times to meet the first condition. The chance of our coin landing heads up is always 50 percent. This meets the second condition. Lastly, the fact that each flip is independent meets the last condition.

Most people, when asked how many heads they would expect, say "**ten out of the twenty**". What they mean is that they expect *about* ten; they really don't expect *exactly* ten heads each time. What comes as a surprise is that the *variation* from ten follows a bell-shaped curve.

Looking at the chart of the binomial associated with our flipping of the quarter, the distribution really looks very similar to the normal distribution. The binomial distribution is just a special kind of normal to take care of proportions, ratios, and percentages.

The idea that "ten out of twenty" occurs only eighteen percent of the time is a surprise to most people. This understanding of the binomial allows us to better interpret percentages and proportions and not be surprised when they are not exact.

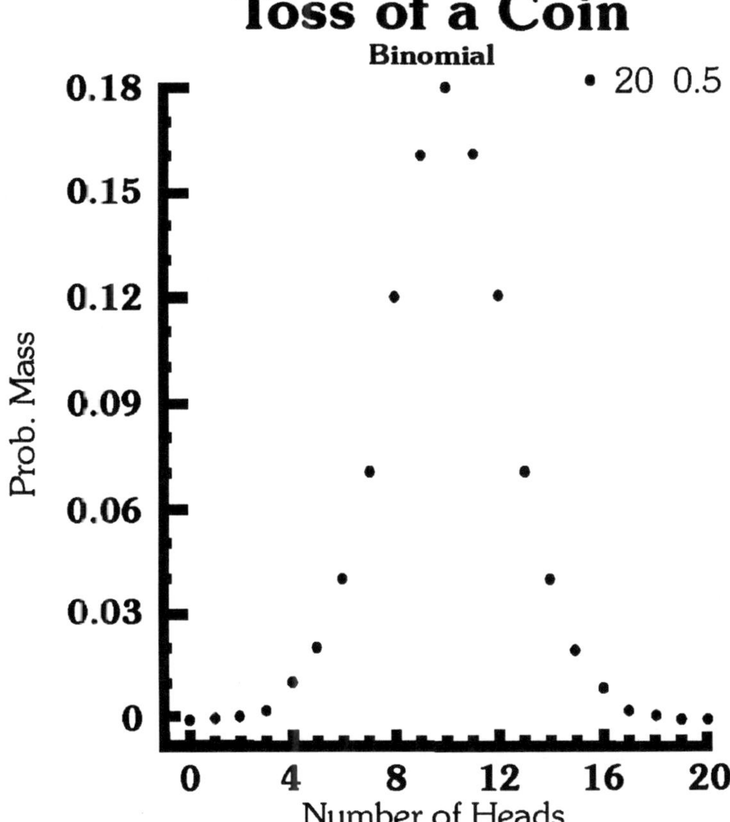

Revisiting our airline baggage problem, we see that the lost bag is very similar to our tossing tails on a coin and that the properly arriving bag is like the flip being heads. The biggest difference is that the starting percentage of lost baggage will be different from the flip of our quarter.

Like the P chart or the NP chart, the attribute control chart also uses the binomial distribution as the basis for its assumptions.

The binomial is a special kind of normal distribution and the normal distribution is used to develop the interpretation rules for variable data. Thus, the out-of-control points of attribute data are found using the same interpretation rules as variable data.

Below is the illustration of the airline's lost bag problem. The illustration shows how the lost bags' distribution would look and similarly how the binomial distribution will appear.

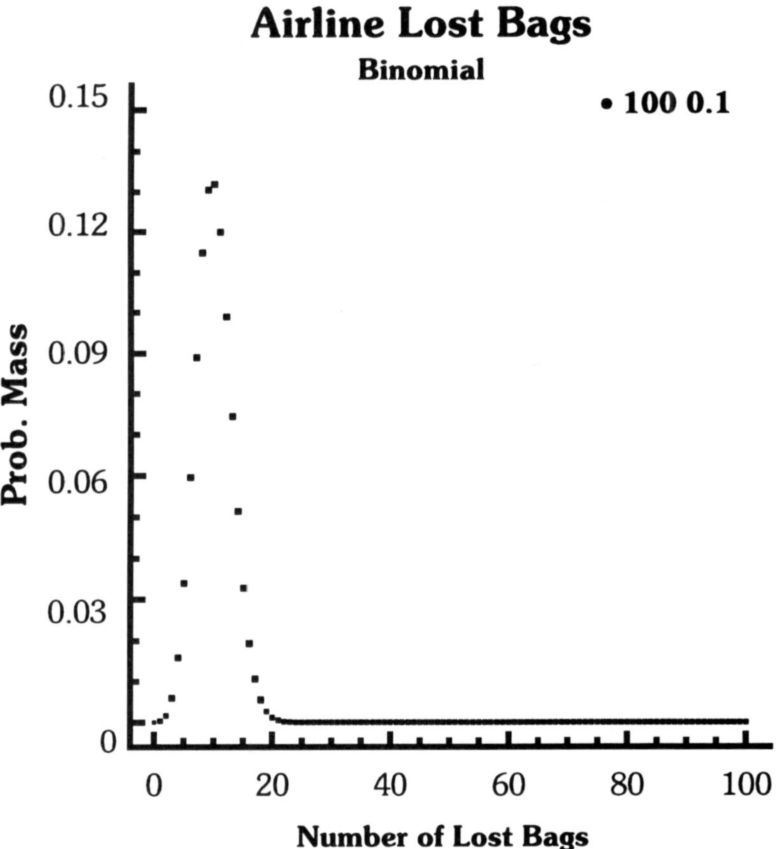

An appreciation of the binomial makes our understanding of the variability of the lost bags on our airline flight very clear.

How is the P Chart Standard Deviation Calculated?

From our explanation of the binomial, the standard deviation formula is $\sqrt{N\ \bar{P}(1 - \bar{P})}$. This formula is designed to show the variability of the number of occurrences. This formula must be rearranged for use in a P chart with proportions, ratios, and percentages. This will be accomplished by dividing the count standard deviation from above by the number of samples in the group.

Standard Deviation of the percentage defective is required and is calculated as follows:

$$\frac{\text{Standard Dev of \# Def}}{\text{Sample Size}}$$

The actual symbols are now inserted in this formula as follows:

$$= \frac{\sqrt{N\ \bar{P}(1 - \bar{P})}}{N}$$

The N is then replaced by the square root of N multiplied by the square root of N. This replacement simply results in the same number.

$$= \frac{\sqrt{N\ \bar{P}(1 - \bar{P})}}{\sqrt{N}\ \sqrt{N}}$$

Next, the square root of N is rearranged in the top portion of the formula.

$$= \frac{\sqrt{N}\ \sqrt{\bar{P}(1 - \bar{P})}}{\sqrt{N}\ \sqrt{N}}$$

The top square root of N is divided by the bottom square root of N resulting in *one* which allows both to be dropped.

$$= \frac{\sqrt{\bar{P}(1 - \bar{P})}}{\sqrt{N}}$$

Thus the **P Chart** uses $\frac{\sqrt{\bar{P}\ (1 - \bar{P})}}{\sqrt{N}}$. Notice the formula here uses the square root of N where in our first P chart the formula used the square root of the average of N. The average technique is allowed as long as the difference between the sizes of the groups is small. If the difference between the sizes is more than twenty to twenty-five percent, then the exact formula should be used as shown here.

What do we do when there are large sample size differences in a subgroup??
If the sizes of the samples in the subgroups vary significantly, then the control limits must be adjusted for each different subgroup. A subgroup size variation of twenty to twenty-five percent from one subgroup to another is considered large.

When the control limits are adjusted from group to group, they shrink as the size increases because our precision increases as the number of data points increases.

Note in the chart below that the control limits spread and shrink. The spreading control limits are from fewer samples in the subgroup because we are less certain of the precision of the subgroup's P value. The collapsing control limits are from increasing the number of samples in the subgroup and the resulting increased P value precision.

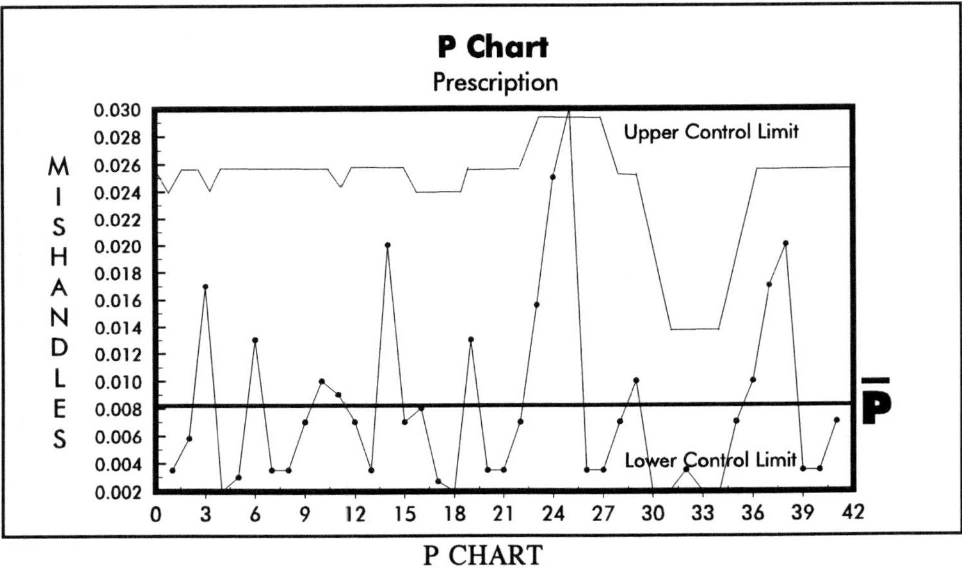

P CHART

Let's return to our airline problem and see how a large shift impacts the control limit calculations. The control limits for the changing sample size in the subgroup are shown for our airline baggage problem.

Should the total number of bags handled from day to day vary as in the example below, then the control limits will have to be adjusted to reflect the radically changing sample size. Notice how the Saturday flight's total number of bags has jumped to 500. This jump is well above our twenty to twenty-five percent threshold so the control limits must be adjusted. This adjustment is not a recalibration of the control limits, but simply a reflection of more precision.

SUBGROUP	Mon	Tues	Wed	Thurs	Fri	Sat
NP	2	3	4	3	3	3
N	200	200	200	150	150	500

Our P values will still be calculated using the same formula.

SUBGROUP	Mon	Tues	Wed	Thurs	Fri	Sat
NP	2	3	4	3	3	3
N	200	200	200	150	150	500
P	.01	.015	.02	.02	.02	.006

Our change comes in the control limit calculation. Let's take a look at these slightly enhanced formulas.

P Charts

P Charts measure the proportion of units non-conforming. The P chart is a special case of the NP chart.

DATA

$$P = \frac{NP}{N}$$

where N equals number of items checked and NP equals the number of items failing.

MONITOR

$$\bar{P} = \frac{NP_1 + NP_2 + NP_3 + \ldots NP_K}{N_1 + N_2 + N_3 \ldots N_K}$$

where K is the number of subgroups studied

$$UCL_P = \bar{P} + 3\,\frac{\sqrt{\bar{P}(1-\bar{P})}}{\sqrt{N}}$$

$$LCL_P = \bar{P} - 3\,\frac{\sqrt{\bar{P}(1-\bar{P})}}{\sqrt{N}}$$

Theoretically, when the sample size changes the control limits should be recalculated. When the N value varies within a range of ±20-25% the use of $\bar{N}$ is adequate for our purposes.

Returning to our airline exercise, the variation of N is large. These large differences in sample size cause our P chart example to have a varying control limit.

SUBGROUP	Mon	Tues	Wed	Thurs	Fri	Sat
NP	2	3	4	3	3	3
N	200	200	200	150	150	500
P	.01	.015	.02	.02	.02	.006

For the airline example, the control limit calculations follow. The first formula is the upper control limit.

Upper Control Limit

$$UCL_P = \bar{P} + 3\,\frac{\sqrt{\bar{P}(1-\bar{P})}}{\sqrt{N}}$$

The upper control limit formula now must be solved with our airline baggage data. The following allows us to compare the two methods for computing the P control limits

With N_{Ave}, our first solution is achieved by using the average N.

$$UCL = .0129 + 3\,\frac{\sqrt{(.0129)(1-.0129)}}{\sqrt{1400 / 6}}$$

$$= .0129 + \frac{3(.1128)}{15.3}$$

$$= .0129 + .022$$

$$= .035$$

With N Now the second solution is computed by using the N for each subgroup. This example uses the N of 500 bags for Saturday.

$$UCL = .0129 + 3\,\frac{\sqrt{(.0129)(1-.0129)}}{\sqrt{500}}$$

$$= .0129 + \frac{3(.1128)}{22.4}$$

$$= .0129 + .0151$$

$$= .0280$$

Lower Control Limit

The next formula computes the lower limit for our example.

$$LCL_P = \bar{P} - 3\,\frac{\sqrt{\bar{P}(1-\bar{P})}}{\sqrt{N}}$$

The lower control limit formula now must be solved with our airline baggage data.

With N_{Ave}, our first solution is achieved by using the average N.

$$LCL = .0129 - 3\,\frac{\sqrt{(.0129)(1-.0129)}}{\sqrt{1400\,/\,6}} = .0129 - \frac{3(.1128)}{15.3}$$

$$= .0129 - .022$$

$$= .0$$

With N Now the second solution is computed by using the N for each subgroup. This example uses the N of 500 bags for Saturday.

$$LCL = .0129 - 3\,\frac{\sqrt{(.0129)(1-.0129)}}{\sqrt{500}}$$

$$= .0129 - \frac{3(.1128)}{22.4}$$

$$= .0129 - .0151$$

$$= .0$$

The tables below show the results of the P chart

SUBGROUP	Mon	Tues	Wed	Thurs	Fri	Sat
NP	2	3	4	3	3	3
N	200	200	200	150	150	500
P	.01	.015	.02	.02	.02	.006
Lower Control Limit	0	0	0	0	0	0
Upper Control Limit	.037	.037	.037	.041	.041	.028

Notice that an extremely large number of bags are processed on Saturday as compared to the other days. Therefore, the control limits are radically tighter.

Now the completed P chart for our example is shown below.

When the probability is low and the opportunities are high, the P chart alarm system becomes unworkable. The problem is caused by the standard deviation calculation. Let's take a look at the calculation that follows: $\frac{\sqrt{\bar{P}(1 - \bar{P})}}{\sqrt{N}}$.

When N is very large, it may be unknown. (How many *opportunities* are there for pinholes in a sheet of paper?) $\bar{P}$ may be impossible to calculate. (What is the ratio of *actual* pinholes to *possible* pinholes?)

The solution is an approximation that simplifies the standard deviation formula above. If $\bar{P}$ is almost zero, then (1-$\bar{P}$) is *almost* (1-0). The formula collapses to $\sqrt{N\bar{P}}$ - the square root of the average number (*not* percentage) defective. In other words, if $\bar{P}$ is very small and N is uncountably large, then a simpler formula can be used which requires neither $\bar{P}$ nor N.

A second problem occurs with a tiny $\bar{P}$ and small sample sizes. The probability of a defect is so small that our expectation is "zero defects" each time we sample. A *single* defect represents change, an out-of-control situation. In these cases, either the SPC is reduced to "CheckMark" SPC or an alternative chart must be devised.

The solution to both of these problems is called a **C Chart**.

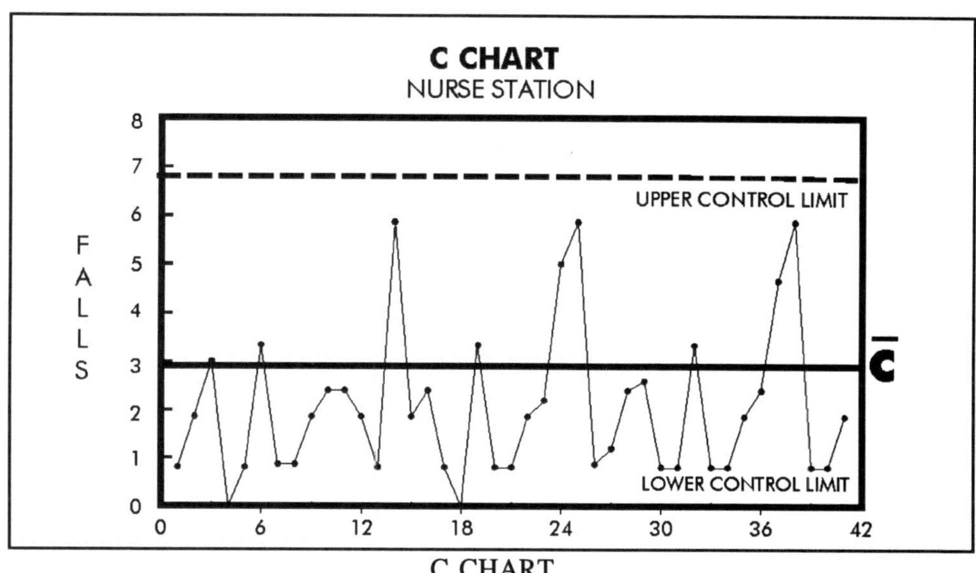

C CHART

The chart above is an example of a C chart. Now let's discuss the calculations that will be used in the C chart.

C Charts

<u>When to Use:</u> The **C Chart** is a special case of the **P Chart** to take care of the following conditions:

1. The probability is low and,
2. The number of opportunities is high.

C Charts measure the frequency or number of non-conformities (defects) where C equals the number of defects.

C equals a count of the number of occurrences

$$\bar{C} = \frac{C_1 + C_2 + C_3 + \ldots C_K}{K}$$

where K = the number of subgroups

$$UCL_C = \bar{C} + 3\sqrt{\bar{C}}$$

$$LCL_C = \bar{C} - 3\sqrt{\bar{C}}$$

Let's continue our airline example on a trip from Nashville to Orlando by monitoring another metric. Over the years, the passengers have changed their baggage-carrying habits due to lost bags and delays in baggage claim at the final destination. To compensate for these failures, many passengers carry their bags onto the plane.

A new problem arises with bags falling out of the overhead compartment. The probability of this happening is really quite low, but the number of opportunities for the bag to fall is high. The control limits are so tight that every time a bag falls, the alarm goes off. The number of bags falling from the overhead is our count and C value.

C Chart Data - Number of fallen bags per flight

SUBGROUP	Mon	Tues	Wed	Thurs	Fri	Sat
C	3	2	1	4	3	

C Chart Exercise

SUBGROUP	Mon	Tues	Wed	Thurs	Fri	Sat
C	3	2	1	4	3	

The above counts are now used to compute the control limits for the situation in which the number of falling bags is changing.

$\bar{C} = \frac{3 + 2 + 1 + 4 + 3}{5} = \frac{13}{5} = 2.6$
$UCL = 2.6 + 3\sqrt{2.6} = 7.4$
$LCL = 2.6 - 3\sqrt{2.6} = 0$

Now take these C values and control limits and build the C control chart for the Nashville to Orlando falling bags chart. The completed C chart for our example is shown below.

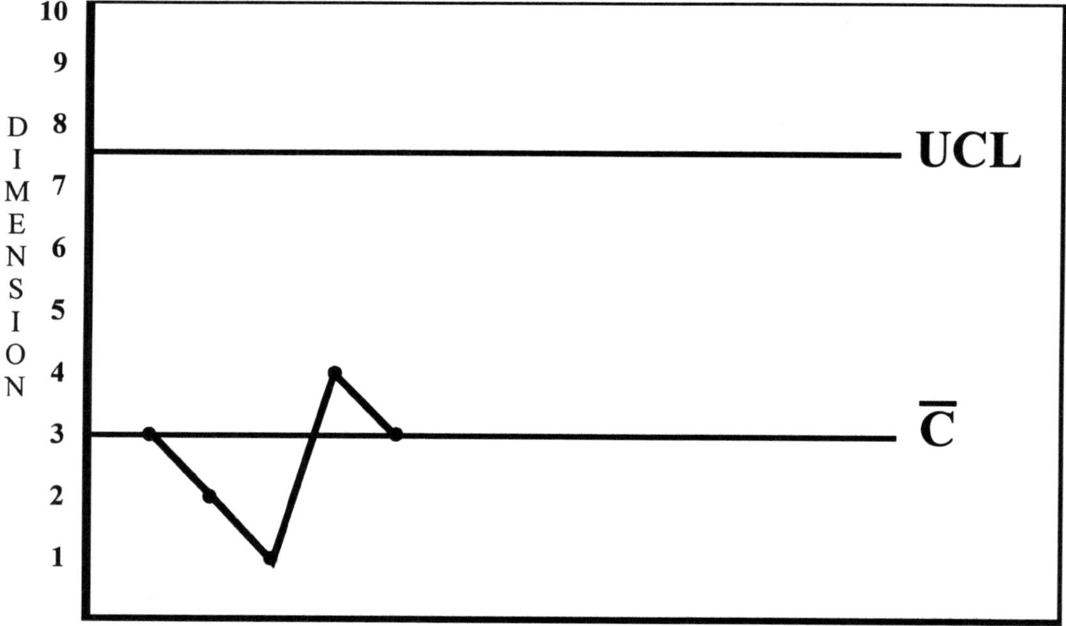

The C chart is read and used like all the other control charts. The alarms from Chapter Eight will be used to interpret shifts and changes. You are probably asking yourself why the C chart uses the same alarms as variable charts and P charts.

When we discussed the variable chart, the alarms were based on the normal distribution. The variable chart alarms use the probabilities of the normal. Most of the values are clustered around the average, with virtually no values falling outside the control limits. The odds of having a value outside the control limits are 0.1 percent.

The P chart, discussed earlier in this chapter, uses a special case of the normal called the binomial distribution. The binomial allows the same variable chart alarms to be used to drive the P system.

Now we must discuss the C chart alarm system. The distribution for the C chart is the **poisson distribution**. The poisson is a special case of the binomial. The poisson is for the situation in which the counts are low and the number of opportunities is great.

Poisson Distribution

The attribute control charts also use the Poisson Distribution as the basis for their assumptions of the U Chart and the C Chart. By using the poisson, which is a special case of the binomial *which is a special case of the normal*, all the control charts use the same fundamental alarm system.

The illustration is of the poisson distribution. The poisson distribution is like the normal and the binomial in its symmetrical shape around the mean. The biggest difference in the poisson and the other distributions is that the poisson stops at zero, and the distribution's bell shape is truncated at zero.

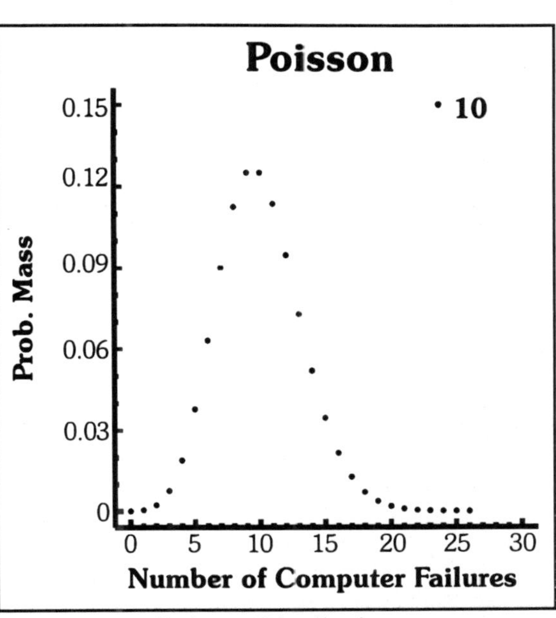

Poisson Distribution

The poisson, being a special case of the normal distribution, allows us to use the same assumptions and alarms that we use for variable data. The formula for the Standard Deviation is the main difference:

Standard Deviation = Square root of the count mean

$$= \sqrt{\bar{C}}$$

Thus the out of control points of attribute data are also the following equations:

Upper Control Limit = Mean + 3 Standard Deviation

$$= \bar{C} + 3\sqrt{\bar{C}}$$

Lower Control Limit = Mean - 3 Standard Deviation

$$= \bar{C} - 3\sqrt{\bar{C}}$$

What if the Unit of Measure Varies?
The C chart is a wonderful tool for monitoring the shifts of counts. A problem arises though when the unit of measure of the C chart varies from group to group. The C chart is designed to monitor counts but only when the unit of measure is consistent.

Another control chart is required to offset the unit of measure problem. This chart is called a U chart. The U stands for the "unit of measure" that the chart deals with. This U chart will also be based on the poisson distribution.

If the flight course is changed and the mileage varies, our unit of measure will not be consistent for each count. Thus the flights cannot be compared to one another consistently. If two bags fall in each of two flights, with one flight covering 500 miles and the second flight covering 1000 miles, the two bags for both flights is not a fair way to monitor the flights.

The count of falling bags need to be related to the distance flown. The first flight should monitor the two bags for a distance of 500 or one bag per thousand miles flown, and the second flight should show its two bags for 1000 miles or two bags per thousand miles flown. Thus the comparison of one bag per thousand miles for the first flight compares fairly when matched against the second flight's two bags per thousand miles flown.

The C indicates the number of bags falling from the overhead. The N represents the number of miles flown. The C and N values are then divided to compute a ratio. The chart below shows the airline flight information.

SUBGROUP	1	2	3	4	5	6
C (Falling Bags)	2	1	3	3		
N (Number Flight Miles)	200	200	200	300		

A new chart is required to fix our varying unit of measure problem. This new control chart is a special type of C chart and is called *U* which stands for the first letter of the *unit of measure*.

What does a U chart look like? The U chart, like all other control charts, has data and alarms. The alarms are once again the center average and the control limits. The chart below is an example of a U chart.

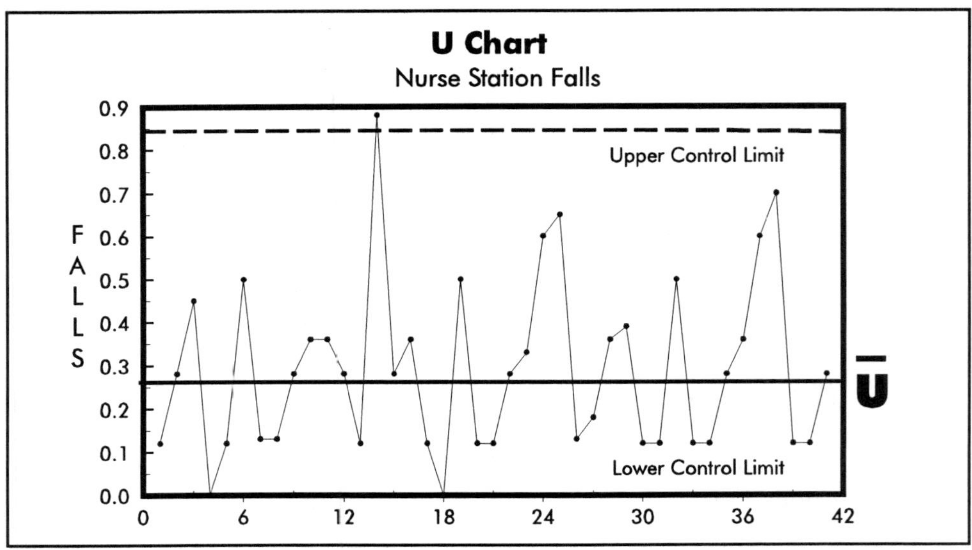

U CHART Falls Per Week

Since the U is a special kind of C, the U is like the C in that both use the poisson distribution for our interpretation rules. Thus the reading of the U chart is like all the other control charts. The details of reading all control charts were covered in Chapter Seven.

U Charts

When to Use: The **U Chart** is a special case of the **C Chart** to take care of inconsistent units of measure.

U Charts measure the proportion of defects by counting the number of non-conformities (defects) (C) and divided by the sample size (N).

C Counts of the number of non-conformities
N Number of samples. An example is the number of yards per roll of fabric.
U Ratio of count of non-conformities to total pieces

$$U = \frac{C}{N}$$

Monitor

$$\bar{U} = \frac{C_1 + C_2 + C_3 + \ldots C_K}{N_1 + N_2 + N_3 + \ldots N_K}$$

The formulas for the **upper and lower control limits** are shown below:

Upper Control Limits

$$UCL_U = \bar{U} + 3\,\frac{\sqrt{\bar{U}}}{\sqrt{\bar{N}}}$$

Lower Control Limits

$$LCL_U = \bar{U} - 3\,\frac{\sqrt{\bar{U}}}{\sqrt{\bar{N}}}$$

Now let's return to the airline problem and compute the U data values for each day. The table below shows the U values for this example.

SUBGROUP	Mon	Tues	Wed	Thurs	Fri	Sat
C	2	1	3	3		
N	200	200	200	300		
U	.01	.005	.015	.01		

U Chart Example Control Limits

SUBGROUP	Mon	Tues	Wed	Thurs	Fri	Sat
C	2	1	3	3		
N	200	200	200	300		
U	.01	.005	.015	.01		

The alarm system is now calculated from the measurements above. These alarms are shown in the table below:

$\bar{U} = \dfrac{2+1+3+3}{200+200+200+300} = \dfrac{9}{900} = .01$
$UCL = .01 + 3\dfrac{\sqrt{.01}}{\sqrt{225}} = .01 + 3\dfrac{.1}{15} = .01 + .02 = .03$
$LCL = .01 - 3\dfrac{\sqrt{.01}}{\sqrt{225}} = .01 - .02 = 0$

Now the completed U chart for our example is shown below.

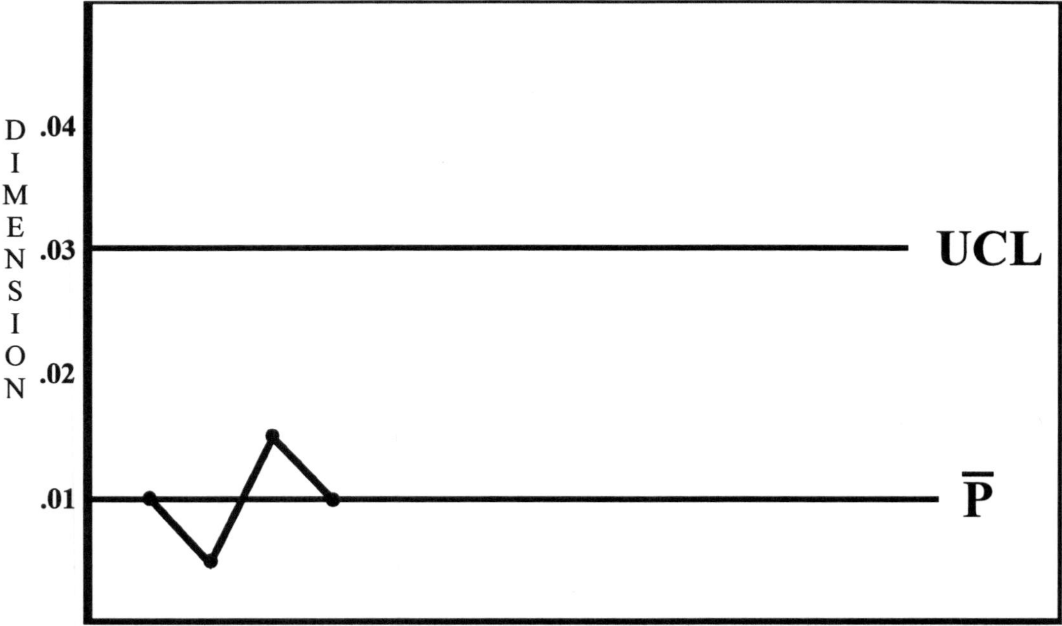

TIME OF MANUFACTURE: CHRONOLOGICAL ORDER

Chapter Summary

This chapter covered the attribute control charts. The calculations, formulas, and where to use each chart were described.

NP Charts measure the frequency of the non-conforming parts. The inspection sample sizes (N) for this chart must be equal.

P Charts measure the proportion of units non-conforming.

C Charts measure the frequency or number of non-conformities (defects) where C equals the number of defects.

U Charts measure the proportion of defects by counting the number of non-conformities (defects) (C) and dividing by the sample size (N).

Remember, these attribute control charts, like variable control charts, are the operator's tool to monitor a particular machine or system in chronological order. Attribute control charts are often confused with management reports, but like all other control charts, they are still a tactical tool.

Attribute control charts include an alarm system with a mechanism for detecting any type of shift or change. The alarms use the same rules that are used for variable control charts.

Now let's move to our final chapter to cover capability studies. The capability study is the companion of the control chart forming the Statistical Process Control (SPC) team. Both tools must be used brilliantly to provide both tactical and strategic players the information that they need.

Chapter Ten: Capability Analysis

Making the Analysis Team Support Strategic Decision-Making?

Throughout this book we have discussed the importance of Division of Labor in both decision-making, goal setting, and proper analysis. This Division of Labor clarifies the difference between strategic and tactical issues. Let's briefly revisit these issues prior to discussing our strategic analysis called capability studies.

Tactical execution is required for strategic decisions to be effective. If operations fails to correctly run the facility, no strategy will work. Our strategic decision-maker must count on our tactical decision-makers to do their job, for only then can the strategic decisions be effective. Many strategic options are available, but they will work only if the operation can run smoothly.

To repeat, management can make decisions only when operations is correctly and consistently running the facility. When this is the case, many options are available to management. When the plant is run incorrectly, we have no way of making a strategic decision that is based on sound logic.

The prediction intervals and confidence intervals become the eyes and ears for effective decisions. The correct set of books must be selected to provide the appropriate information to support the type of decision being made. The correct presentation of information is essential to good, sound decisions. Failure to provide or choose the correct set of books and the correct analysis tool will yield incorrect information.

When strategic and tactical personnel both do their jobs, they form a team. Operations is in deep trouble when management does not do its decision-making job of providing the facility. If the facility (available means) is not capable, we are doomed to having dissatisfied customers and stockholders. If operations does not perform its role of running the facility (available means) correctly and consistently, then how can management assess the impact to the customer?

The correct selection of the analytical tool then becomes of paramount importance such that the tactician and the strategist can work together. Just as strategic and tactical decisions form a team, confidence and prediction intervals also form an essential team for properly prepared information. Both the decision-makers and the proper support information must come together to build a finely honed partnership that allows for good, effective, clear decisions.

In Chapter Two we discussed support for both tactical and strategic decisions and goal setting. Two sets of analysis tools are provided: one for strategic decisions and the other for tactical decisions. The following table summarizes the analytical tools to support tactical and strategic decision-making and goal-setting.

	Tactical	**Strategic**
	Operations ↓ Detect Change ↓ Averages	Management/Customers/Sales ↓ Assess Capability ↓ Individuals
Central Tendency	**Ave (Ave)** =	**Ave (Ind)**
Variability	$\mathbf{STD_{Ave}}$	$\mathbf{STD_{Ind}}$
	$\mathbf{STD_{Ave}} = \frac{\mathbf{STD_{Ind}}}{\sqrt{\mathbf{n}}}$ Where n = subgroup size	
Distribution	Central Limit Theorem Normalizes Data	Histogram
	Confidence Intervals $X_i = X_{Ave} \pm z * \frac{Std_{Ind}}{\sqrt{n}}$	Prediction Intervals $X_i = X_{Ave} \pm z * Std_{Ind}$

All reports must assure that the proper set of books matches the proper decision type - either strategic or tactical. No one analysis tool can support both types. Both sets of books must complement each other and work as a team. Since this chapter deals with capability studies (strategic analysis) then prediction intervals or individual readings must be used.

Throughout this book, the methods to effectively control a process and product have been discussed. As a review, the following are the steps:

- Calculate initial control limits
- Establish that the process is in control
- Continually survey the control charts for alarms (out-of-control conditions)

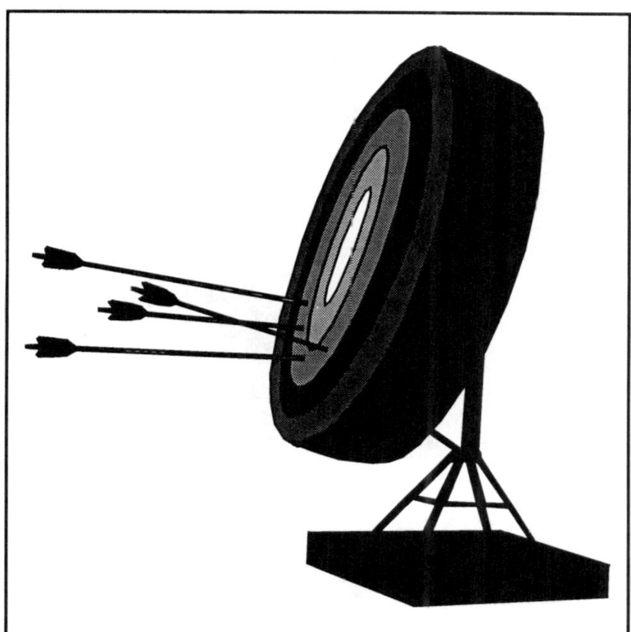

Establish Process Control

As the drawing illustrates, the first essential component is establishing a consistent pattern. My favorite sport as a boy was playing third base in baseball. My skills with the glove were always flawless. My teammates called me "the vacuum cleaner" because nothing ever went past me. However, this didn't guarantee an out for our team because I had a shotgun arm to make the throw to first base. To explain my problem a little, when I made my throw, the first baseman and I were always in a panic. We both knew that with my release, the ball would have tremendous speed, but nobody, including me, knew where the ball was going. Sometimes it sailed perfectly to first for the out, at other times it landed in the dirt half way to first, and sometimes it flew straight into the stands. You can see the importance of consistency.

Now with the process in control we are ready to discuss the methods to effectively assess the capability of the process and product. Once again our initial steps are the following:

- Calculate initial control limits
- Establish that the process is in control
- Continually survey the control charts for alarms (out-of-control conditions)

And now we're ready to begin the final step:

- *Assess the capability of the in-control process*

Our final step allows the strategic decision-makers to assure that the facility is truly capable of making the product the customer wants.

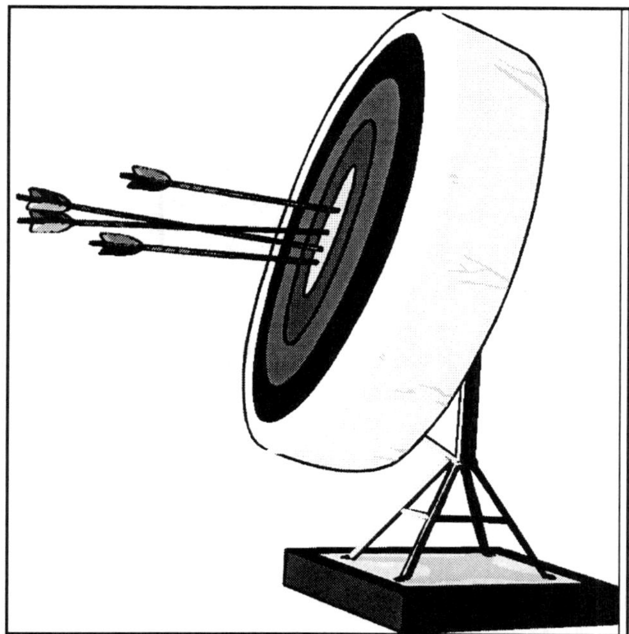

Continuing Improvement
...Through Control and Capability

As in the drawing, a consistent product and a capability study show us what and where the improvements to our product and process must be made. The strategic decision-makers can now use this knowledge to achieve a bulls eye every time. This bulls eye will equate to improved product and processes.

Now lets recap the control charts and see how they must work to make the capability study accurate.

How does Control Fit With Capability?

For a proper capability study, your homework must be completed. This homework is the proper building and grouping of your data and is fundamental to obtaining a proper understanding of the capability of your machine or process. In baseball, much hard work is expended before the first game is played. Spring training and all our practice sessions must be complete so that the season can begin and achieve the results that a baseball team desires: ***WINS!!*** In a continuous process, our homework is tactical execution which must be monitored and pushed to a level of brilliance.

The tactical tools to assess that the fundamentals are performed correctly are the control chart and the Walkabout™. The control chart is the tool to assess the consistency of the product or process. The Walkabout™ is the tool to monitor the correct running of the machine. The Walkabout™ and the control charts must work together to state that the machine is running correctly and consistently. These are the tactical tools used by the operator to assure the strategic decision-makers that all is ready for our capability study. These tools have been covered in-depth in the prior chapters of this book.

Now we are ready to begin. Then and only then, after the tactical force's job is complete, can we state that we are ready to assess the capability of the process. Now our strategic decision-makers can take over.

The tactical and strategic forces must work together as a team to provide good information just as a pitcher and a catcher must work together. The catcher calls a good mix of pitches to keep the batter off guard, and the pitcher then must deliver these pitches where the catcher calls for them. Finally, the catcher must catch the pitch that is delivered.

If either player works out of harmony with the other, then disaster is brewing. Should the pitcher ignore the catcher's call for a fast ball and deliver a curve ball, the ball is a very likely to get away from the catcher. Even if the pitch is a strike leading to a strikeout, the batter can still get to first because of the wild pitch.

For comparison's sake, the control chart is the tactical tool that monitors execution of the plan, just as the pitcher is the tactical person that must deliver the correct pitches in the called for location.

What does Control Mean?

Control means consistency. When I was younger and lived in the small town of Alexander City, Alabama, I coached a youth baseball team. My coaching approach was unique in that we built a team of very specialized players. Early on in my first player draft (you can see we took this very seriously) one very young kid named Greg wanted to be a pitcher but was being overlooked by the other teams. When this kid pitched, he wasn't the fastest nor the most overpowering. He didn't have the sharpest curve ball nor the largest selection of pitches. What this kid had was consistency - every pitch was exactly a strike!

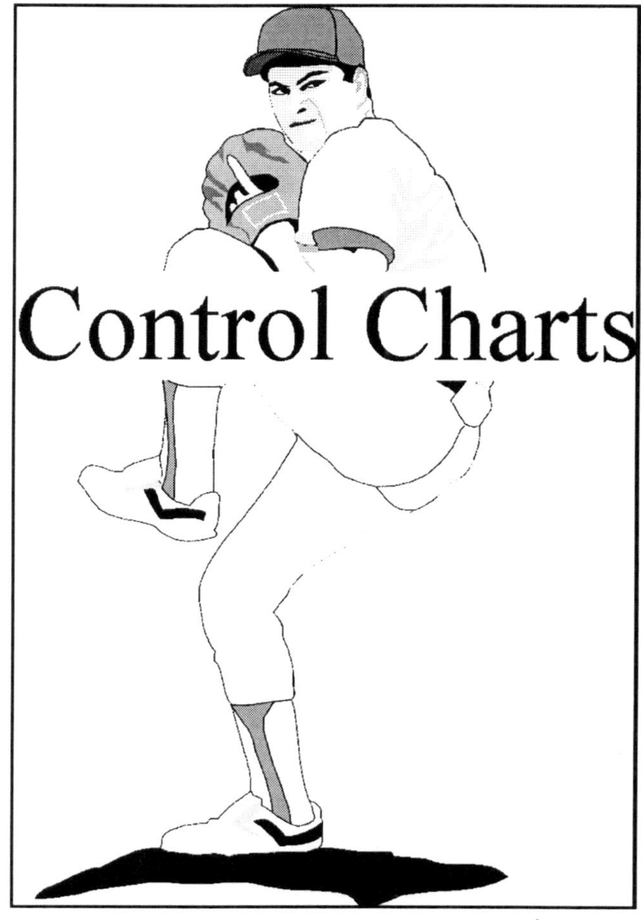

I drafted him and planned to use him as our relief pitcher. We worked with him to assure that he made a strike on every pitch. As the team played together, Greg knew his role and was never disappointed because he didn't start. He knew he would be there when the team needed him.

When our team was in trouble and needed to stop an opposing team's rally, we knew who our relief pitcher was. When my players saw Greg warm up, they would cheer because they knew we were about throw cold water on the other team. In fact, they began to call him "the fireman." We all learned that Greg was consistent and always threw strikes which put pressure on the opposing team.

This predictability won many a game for our team. *In baseball, this type of pitcher is called a control pitcher.* He knows where the ball is going to be placed and positioned each time he throws the ball.

Consistency means *predictability*, and consistency and predictability are synonyms for the word **control**. *Capability* says we are predicting the future from the past. Should we attempt to make a prediction from something that is unpredictable, then we must question our results.

Out-of-Control means the process is not consistent and is unpredictable. If the process is capable but is out-of-control, the future is not predictable or represented well based on the past and present data. Similarly, if the process is not capable and is out-of-control, bad product is not necessarily inevitable.

In Control means the process is consistent and predictable. If the process is capable and is in control, the future is predictable because of the consistent nature of the past. Our capability is represented well based on this past and present information. Similarly, if the process is not capable and is in control, bad product is definitely coming our way.

What about Process Metrics?

Don't forget the Basics from the First Process Management Principle. The first principle says the process must be improved for the product to improve. We must not forget that the ***P*** in SPC stands for *process*. The process metrics must be run correctly and consistently.

The maintenance and compliance of the facility (housekeeping) is fundamental to the correct and consistent running of the facility. No statistics, analysis, expert systems, experimentation, automation, or computerized system can overcome housekeeping lapses. An automated facility that is not maintained will produce huge problems, inefficiencies, and defects. Automation demands good housekeeping and manufacturing practices. Housekeeping demands a continuous and ongoing ***dedication*** to the maintenance of the facility. Good housekeeping is an absolute must to achieve improvement. A clear understanding of the linkage between the principles of process management, improvement, quality, customer satisfaction, and the long term viability of the facility must be understood.

Additionally, an ongoing program to reinforce the importance should be set up. The ongoing program should resemble regular safety meetings. These operator housekeeping meetings should start with weekly sessions that include top and middle management attendance. Once the importance has been established, the weekly frequency could be revised to a monthly session with operators and line supervision only in attendance.

Walkabouts™ are the focal point for assessing the compliance of the housekeeping effort. The Walkabout™ provides an easy way to document for an operator one machine's running parameters. A training class to explain a Walkabout™ must be conducted.

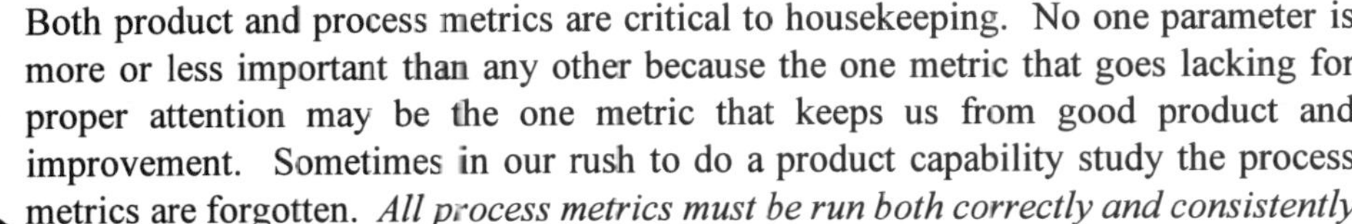

Both product and process metrics are critical to housekeeping. No one parameter is more or less important than any other because the one metric that goes lacking for proper attention may be the one metric that keeps us from good product and improvement. Sometimes in our rush to do a product capability study the process metrics are forgotten. *All process metrics must be run both correctly and consistently for the product capability study to be valid.*

All associates must understand what we are trying to achieve through the Walkabout™, and they must use the Walkabout™ to RUN the machine. We will expect to have a Walkabout™ for ***each*** machine and product. The Walkabout™ gives us a tool to not only run the facility but to assess compliance as well.

Keeping the Facility running Correctly The Walkabout™ is the tool that allows for a clear documentation of the process. It must be a dynamic tool that is used by all associates. The Walkabout™ gives us a benchmark to begin our changes and allows all associates to understand the process. Good housekeeping is of prime importance to all associates. When we deviate, we have a clear procedure for documenting change. Depending on the magnitude and impact of the change, we will need to report the change up and down the decision ladder.

Grandma used the Walkabout™ to explain to my mother the making of southern biscuits, step by step. The Walkabout™ first became a tool to document how Grandma made the biscuits. Then Grandma used the completed Walkabout™ to teach my mother how to make the biscuits. Finally, the Walkabout™ became Mama's tool to check to see that everything was done properly. ***Grandma and Mama both kept good housekeeping practices.***

Walkabout™ for Compliance: The Walkabout™ now is the operational tool to assure that the operator knows and can recall how to perform a particular procedure. We can use the Walkabout™ as an operational tool to document the system's procedure so that we can measure our product and process.

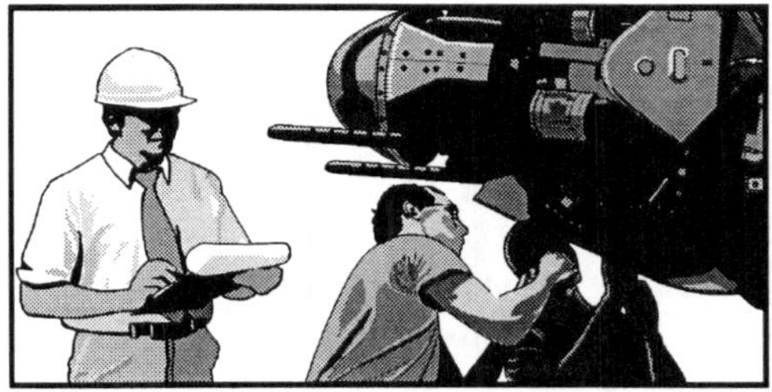

The Walkabout™ also provides a medium for all management levels to conduct their own audit of the facility's metrics compliance. Management should periodically make an audit of each machine. This would simply require the manager to take the operator and the Walkabout™ around each section of the machine. ***The manager and the operator together*** can then test the compliance of the machine's metrics. The manager and the operator should discuss why the metric is there. The operator should be able to explain why each metric is important. If he does not understand the metric or its importance, then training of the operator's process Knowledge Base can be initiated.

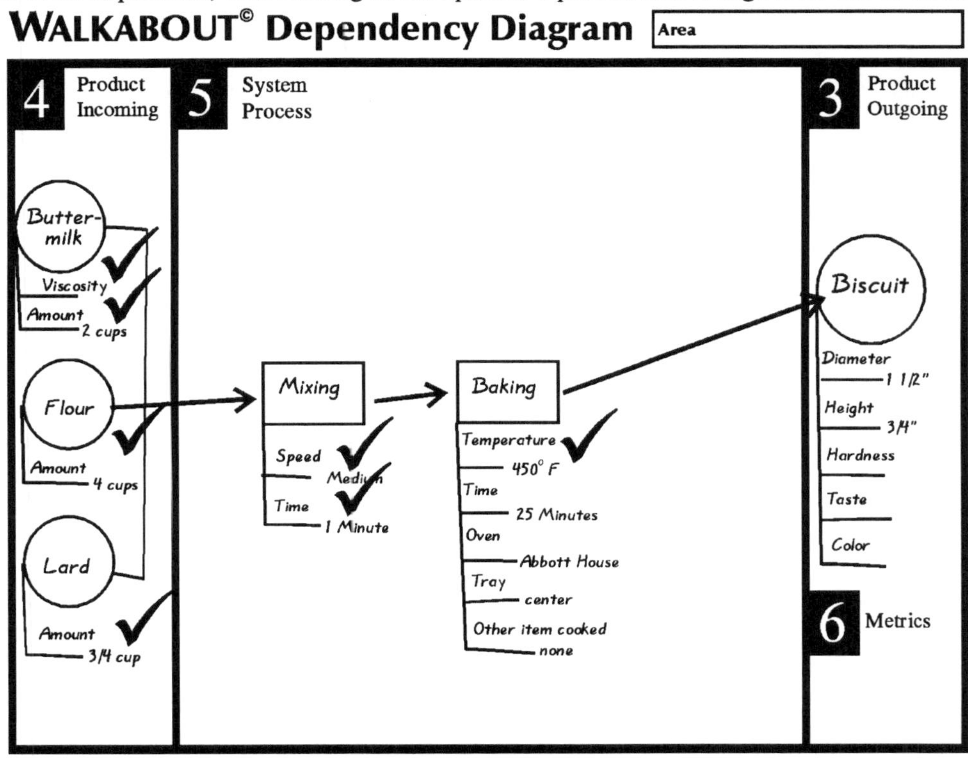

With a process Knowledge Base, we now have a benchmark to begin an improvement program, Statistical Process Control, and other efforts.

Keeping the Facility Running Consistently. Statistical Process Control Charts provide a tool to monitor the consistency of the process. Naturally, then, they also provide a means for an operator to detect any shifts in the consistency of the process or product.

Since this is a tool to assist in the consistent running of a machine, all reporting must be for one particular machine. A machine will have multiple process and product metrics for our monitoring, but each process and product metric will have its own control chart.

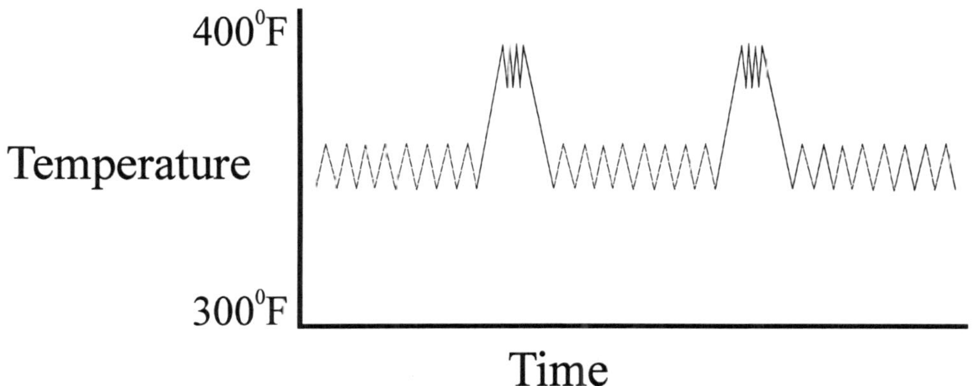

Control charts become our tool for assessing over time if we are correctly running the facility. All deviation from the correct and consistent running of the machine must be reported and assessed for impact on the customer.

In addition to documenting the change, all knowledge of **why** the change occurred must also be documented and reported. This information may allow our strategic decision-makers to replicate the cause to other machines.

Remember, Process Metrics must be in control also.

The process metrics are vital to an accurate capability study. The control of process metrics assures that the process is stable and producing a consistent product. Thus, the capability study requires that all process metrics be run both correctly and consistently. The correctness of the process and product is validated by the Walkabout™. The consistency of the process and product is validated by the control chart.

Where Does Capability Fit in the SPC Model?

Just as the members of a baseball team must work together, those of us in manufacturing must work together to make tactical and strategic decisions like a well-oiled team. Control charts provide the proper information to support tactical decisions, and capability studies provide the proper information to support strategic decisions. Each member of a baseball team must play his position well; similarly, each member of a decision-making team must have access to correct and appropriate information.

The Process Control Method
The overall objective of Statistical Process Control (SPC) methods is to prevent product or performance defects.

How Many Defects Should We Tolerate?
If the aim of process control is to prevent defective product, we must predict the percentage of defects from the controlled process. In turn, this requires that we know about the distribution of values from the process itself.

However, our data at this point is the averages of samples drawn from the process for the tactical purpose of running the facility - not the process values themselves. We can use the same sample data that appears in our control chart to estimate the location and spread of the process distribution and thus the proportion of product that falls outside of the customer's specifications. To this end, we must analyze the values with strategic tools called **prediction limits** which use the individual readings. How the individual values are used will be covered later in this chapter.

What Is Acceptable Quality? What is the acceptable quality for the product in our organization? A quality program can be set up with two philosophies. One approach is detection and the other is prevention. Using a detection philosophy our organization tolerates defects and must rework the products that are defective. Using a preventative philosophy, we are working toward controlling and eliminating defects.

In the preface of this book we have defined quality in a way that is extremely broad and encompassing. According to this definition, quality is comprised of three components: performance, cost, and time. Our research, analysis, and decisions must support our goal of delighting our customer (Performance), while reducing our cost (Cost), and providing our product to our customers the moment they want it (Time). We must view this goal as a never-ending journey that continues to improve all three components because no one component is more important than the others. All three must be viewed with equal importance and balance.

The balance of the three components of quality is crucial to an accurate capability study. Using a detection quality philosophy keeps bad product from the customer, but our cost will be higher because of the inherent rework and scrap that this philosophy tolerates. To keep a balance among the three components of quality, our mode of operation must be preventive to stop detrimental change before it becomes a problem.

Once the capability study is complete, then the control charts become the alarm system to alert us to change in the product or process. Remember, once a change is detected management must be alerted so that a new capability study can be developed. A change can be either an improvement or a deterioration, thus SPC forms a powerful team for improvements.

The capability study is the strategic component of SPC. Capability studies allow us to understand how we are meeting our quality objectives.

What are the Steps for the Capability Study?

The following are the steps required to set up the capability study. Each step in the journey to an accurate capability study will be covered in detail in this chapter. The steps are:

1) Determine which data points to include
2) Calculate standard deviation
3) Determine the specifications
4) Test for normality

Determining Which Data Points to Include in the Capability Study

In collecting information for a capability study, many people simply pull all the data readings for the past year. They assume that the more readings we have, the better the capability study must be. In Chapter One, though, two analysis flaws were uncovered. Because improper technique is the first analysis flaw, this chapter will focus on assuring that proper techniques are applied to our analysis. For proper numerical analysis, our technique must monitor each of the following:

- Central Tendency
- Variability
- Distribution

Failure to monitor any one of the three will have dire consequences.

Simply having proper technique of a year's worth of data is not enough though. Our groups must also form one universe (homogeneity) based on physical similarity and time of manufacture.

We can assure proper physical groups by employing common sense. The time-series groups are formed based on the analytical tool called **control charts**.

For us to do the detailed analysis that will lead to accurate reporting, a great deal of data gathering will be required. To adequately handle the detailed reporting, all associates must participate in the data gathering effort. To support proper group formation, a bottom up strategy of data reporting is required. This bottom up group formation will test for the time-series groups. The control chart is the tactical tool to assess the time-series groups.

What are the capability requirements for selecting the data points?
Each machine's product must be kept separate by time frame such that we can identify homogeneous groups. The only alternative to assuring correct formation is to assure the following:

1. Make sure each individual machine is monitored.
2. Use control charts, the SPC tool to identify when the time-series homogenous group has shifted.
3. Make sure the monitoring is set at an appropriate time interval for each machine to assure a continuous surveillance. The time interval must be set such that any time-series shift can be detected.
4. Make sure a detailed investigation is undertaken when a shift has been noted. This investigation must determine the cause. The effect of the cause can be either an improvement or a deterioration.

All our analyses are based on the assurance that both proper technique and groupings have been used. This combination will allow us to provide accurate capability studies and will require a bottom up strategy of data collection and analysis.

The need to create physical groups according to individual machines and differing products should be clear from Chapter Three. The time of manufacture or chronological grouping is a little less obvious and requires the special facilities provided by a control chart. The control chart is our means of being certain that the machine is in a state of ***statistical control***.

What Does Statistical Control Mean?

To define the term **statistical control** let's look at each word. **Statistics** means that we are following a prescribed mathematical procedure to summarize and provide a picture of our numbers. For our purposes, we are using past historical information about a product or process to describe the current and future results of the machine. **Control** means consistent and predictable results are being accomplished. Putting the two words together means that past history is being used to project the future of a consistent and predictable product or process.

We have discussed the idea and use of control limits and control charts. Control limits are the limit of control but they have no relationship to a specification and are generally three sigma limits around the mean. When a process is in statistical control, this means the process is quantitatively consistent and predictable.

A process that is statistically in control is stable; we can perform a capability study. What would happen if we attempted to do a capability study on a process that was out of control?

A capability study must be done when the process is stable.

Let's see an example of why control is crucial to a capability study. The chart in the diagram here shows the band of our product to be very erratic. The results move high and low, and the results are also spread and tight. If this data were plotted on a control chart, the data would be identified as being out of control. These values would make the results of a capability study unpredictable at best and, as we will see later, wrong.

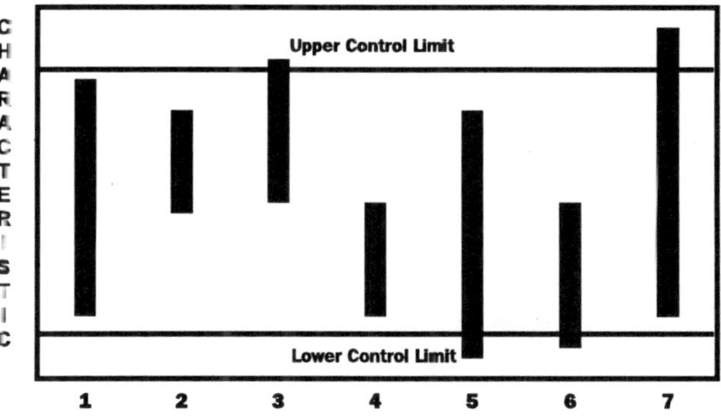

What would happen if a capability study were done on this type of data? To clarify why being in control is so crucial to a capability study, let's analyze a facility and see the control issues and the capability issues.

Understanding Capability by Implementing SPC

Let's return to the example of the candy plant that we used earlier. As a boy growing up in Nashville, Tennessee, my greatest treat was a candy called GooGoo Clusters. These little bits of heaven were and still are made in Nashville by Standard Candy Company. A GooGoo contains the finest ingredients known to man. In the center, the finest marshmallow is surrounded by caramel and the best peanuts grown in southern Alabama and Georgia. The finest chocolate produced in Tennessee or anywhere else in the world covers the entire confection. I am a connoisseur of GooGoos, and being a connoisseur, I know that a GooGoo Clusters weighs roughly 50 grams because, of course, every time I eat GooGoo Clusters, I weigh and inspect the package.

Grand Ambition: My grand ambition and most exciting fantasy is to own Standard Candy Company. Then, I could eat as many GooGoos as I wanted, I could give them to my friends, and I could sell them.

Meeting my fantasy and the demand: In my fantasy, I purchase Standard Candy Company. I decide to build a new state of the art plant in Nashville, Tennessee, and bring in the most competent engineers in the world to design this plant. We hire the best construction company, and we use the best material and equipment in building the plant. Upon completion, the plant will make GooGoo Clusters that weigh about 50 grams.

Customer and Management Expectations: To assure our customers of quality candy, we have an unconditional guarantee which covers taste, freshness, ingredients, and weight. If there are any deviations, the customer will receive an immediate refund. Our purpose here is to assure that no customer receives a GooGoo that is too light. The guarantee states that no GooGoo which weighs less than 45 grams will be shipped. We must also protect the investment of the stockholders, for if GooGoos are manufactured that are too heavy, then we are giving away product and raw material, and this generates no compensation for the company because we are not charging a higher price for the extra ingredients. To protect our shareholders, we have set a specification of no GooGoo being manufactured that weighs more than 55 grams. Our specifications, therefore, protect both of our customers (shareholder and end consumer).

Operations: Before we proceed, I must explain the operations of the facility. Because the facility is state of the art, the engineers have detailed instructions for its correct running. When the facility is correctly and consistently operated, 50 gram GooGoos (or as close to 50 grams as possible) will be produced. The person in charge of operations will be responsible for the correct and consistent running of the facility, and we will hold him accountable.

Why Can't We Do a Capability Study if the Process is Out of Control?

The chart shows the exact amount of product produced for the varying weights. The weights are not exactly 50 grams but as close as possible for this facility. The chart details the actual production by weight for one of the GooGoo machines. In an hour, the machine will produce 3000 GooGoos whose weight will breakdown into the following: 100 GooGoos that weigh 48 grams each, 400 GooGoos that weigh 49 grams each, 2000 GooGoos that weigh 50 grams each, 400 GooGoos that weigh 51 grams each, and 100 GooGoos that weigh 52 grams each. Every GooGoo machine in our facility consistently produces this mixture every hour.

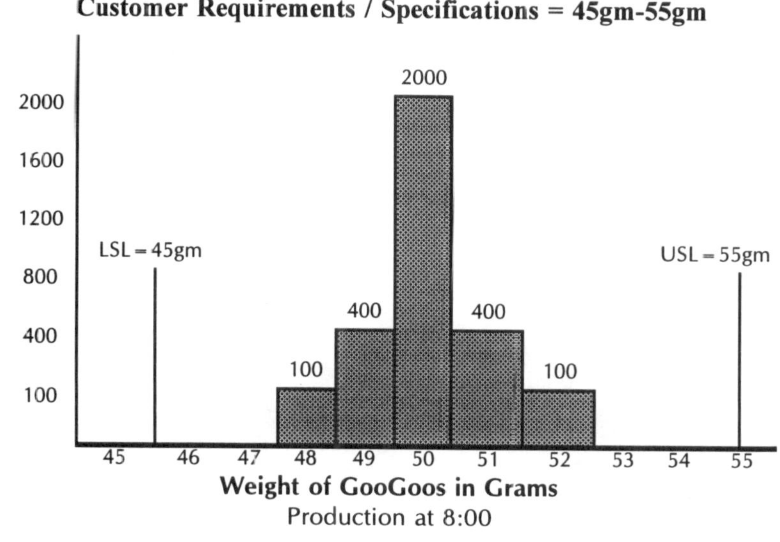

Our process, which is extremely capable, is shown in the graph. We see how the production at 8:00 impacts the customer. No defective product has been produced, for no product depicted in the bars of the chart is above or below our customer specifications.

The quality department issues the revised guidelines that constitute a continuation of the Quality Hell philosophy. The guidelines read, "Whenever a product out of specification is encountered, adjust the machine in the direction that corrects the defects. Five products must be sampled every hour. When any of the five sampled products do not meet the customer's specification, we must adjust the machine. *Only make adjustments when defective products are encountered.*"

Our strategic impact this time is different. We will only adjust the process when defects are encountered. Any drifts in the process are ignored as long as no defects are detected. No normal adjustments will be made until the process begins to get close to the upper or lower specification.

The graphs below chart the production and the allowable drift created from the new procedure.

Customer Requirements / Specifications = 45gm-55gm

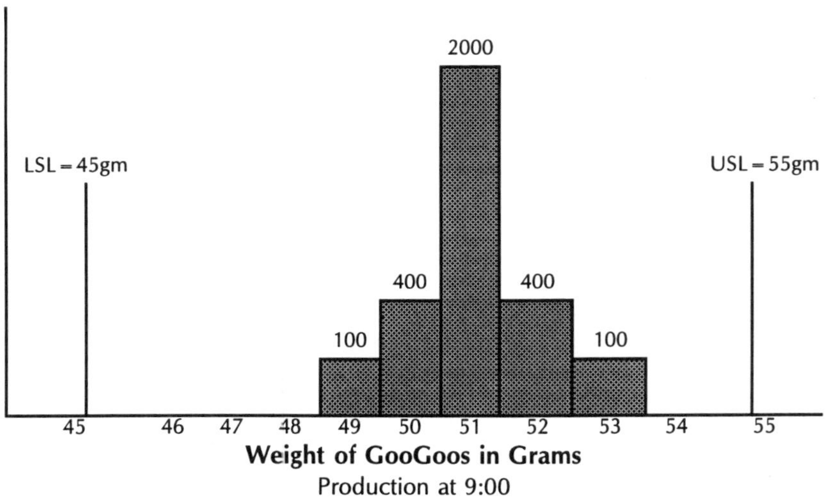

At 9:00, the process has begun to drift upward on the heavy side. The center has drifted from 50 grams to 51 grams.

Customer Requirements / Specifications = 45gm-55gm

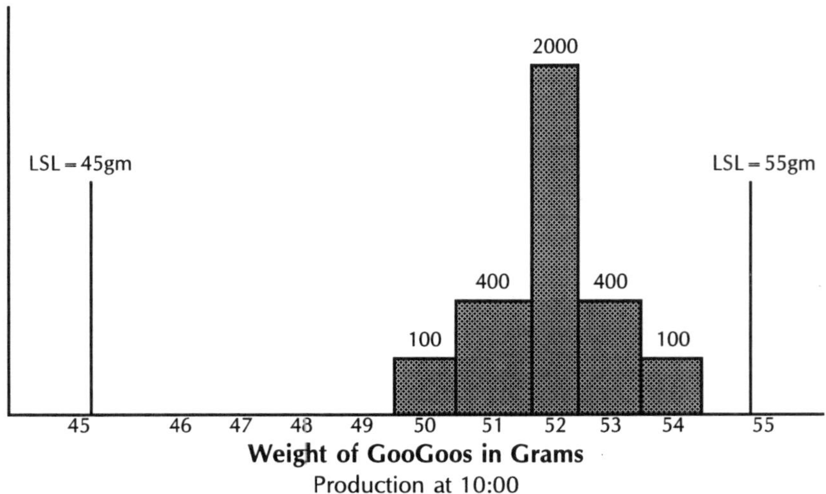

At 10:00, the process has continued its upward drift with no correction. Our center is now at 52 grams.

Customer Requirements / Specifications = 45gm-55gm

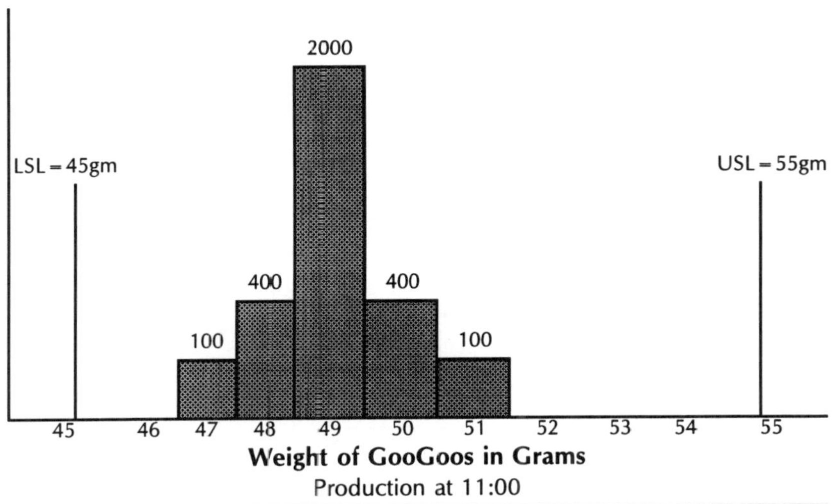

> At 11:00, the process has slid on a downward drift from a center of 52 grams to a center of 49 grams.

Customer Requirements / Specifications = 45gm-55gm

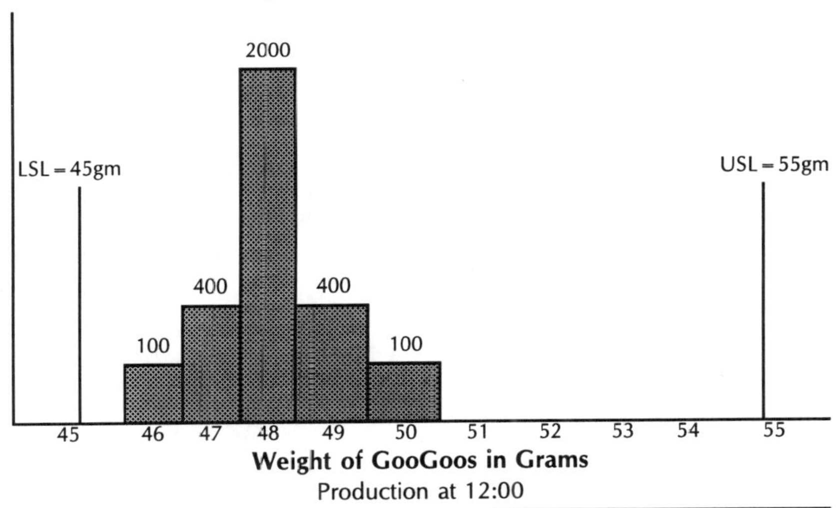

> At 12:00, the process has continued its downward slide to a process center of 48 grams.

At no time between 8:00 and 12:00 were any defects being produced. Thus our operations staff would have had no reason to make adjustments or to feel they were not running the machine correctly.

We will record the results of the products produced between the hours of 8:00 and 12:00 in a set of summarized reports. Notice the products' weight spread after we total all products.

	Units Produced by Weight										
Production	**45**	**46**	**47**	**48**	**49**	**50**	**51**	**52**	**53**	**54**	**55**
8:00				100	400	2000	400	100			
9:00					100	400	2000	400	100		
10:00						100	400	2000	400	100	
11:00			100	400	2000	400	100				
12:00		100	400	2000	400	100					
Total Production		100	500	2500	2900	3000	2900	2500	500	100	

When we run an extremely capable facility using customer specifications, we simply assume that any running parameters within specifications are correct. This assumption could lead us to the above condition in which the process is floating around haphazardly, oscillating within specifications. We have no concern for running the machine as correctly and consistently as possible.

By graphing the summary results of the table above, we see how the customer is impacted. If we used the graph on the right to assess the machine's capability, we would believe, incorrectly, that the machine is only barely capable of producing the product required.

Customer Requirements / Specifications = 45gm-55gm

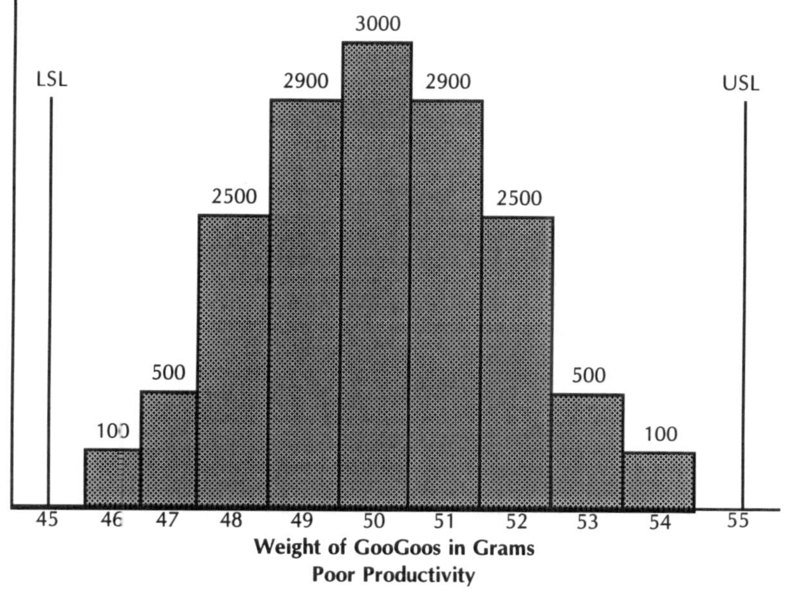

Weight of GooGoos in Grams
Poor Productivity

If we violate the second principle by making machine adjustments based on customer specifications on an extremely capable process, we will make this process appear less capable than the machine really is. We must conclude that we are not running this machine correctly and consistently.

If our 8:00 to 12:00 production had been run as correctly and consistently as possible, our production of GooGoos would have been as the table below shows.

	Units Produced by Weight										
Production	**45**	**46**	**47**	**48**	**49**	**50**	**51**	**52**	**53**	**54**	**55**
8:00				100	400	2000	400	100			
9:00				100	400	2000	400	100			
10:00				100	400	2000	400	100			
11:00				100	400	2000	400	100			
12:00				100	400	2000	400	100			
Total Production				500	2000	10000	2000	500			

The results of the correctly run facility can be graphed. We'll now compare that to the results of the incorrectly run facility from the previous page.

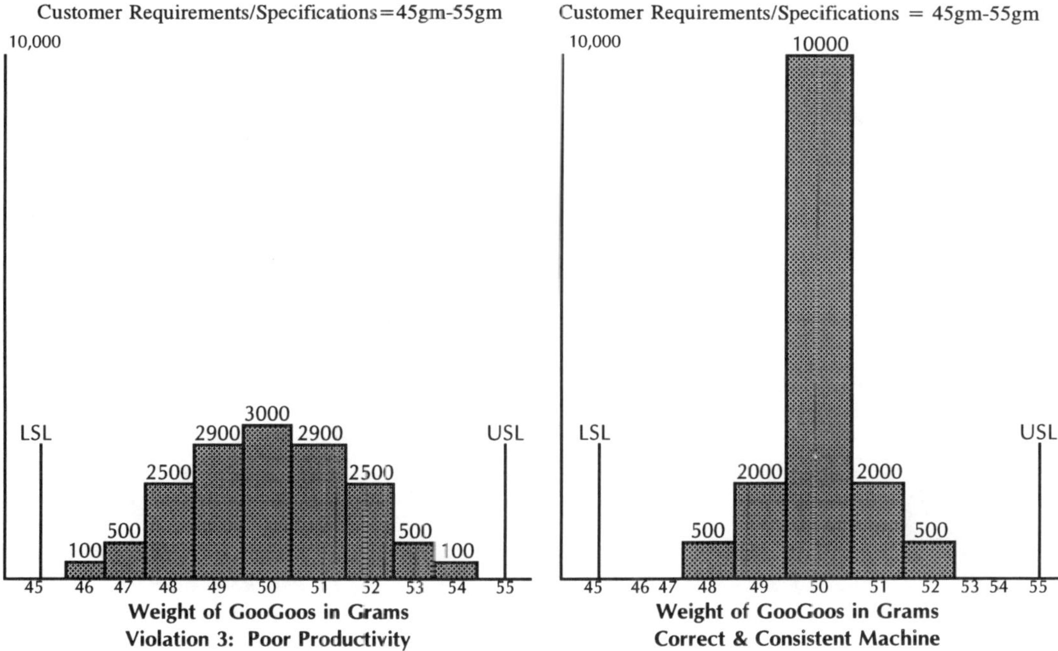

Weight of GooGoos in Grams
Violation 3: Poor Productivity

Weight of GooGoos in Grams
Correct & Consistent Machine

When we note the widely different distributions, we develop two completely different impressions of the same machine. Clearly this could lead to major confusion and potentially inappropriate actions. We must make sure that any decision or action is based on a machine that is run correctly and consistently. Therefore, a clear understanding of the Second Principle of Process Management is the key to making sure that the machine is run properly at all times. Obviously, our impression of the capability of this machine is radically altered depending on which graph we view. We should always base strategic decisions on the validated assumptions of the correct and consistent running of the machine.

From our candy example, the potential for erroneous results is huge. Thus a control chart is essential for an accurate capability study. Not only could our capability study be incorrect, but improper preparation can cause poor and flawed decisions to be made. Thus statistical control is essential to an accurate capability study.

How do we use Statistical Control in a Capability Study??

Should we use bands of data that are out of control (non-homogeneous), this would be like mixing apples and oranges. Remember, out-of-control means inconsistent and also improper grouping. For us, the result is very camouflaged and confused. No decision from this kind of analysis would be valid. Let see an example where the values are improperly grouped.

The table below is the candy weights from one candy machine. Notice that the candy process runs along for time zone A consistently for about two days, and then shifts in time zone B. In time zone B the weight has dramatically shifted down for a period of two hours. Finally in time zone C the candy weight moves back up to its consistent weight. Time zones A and C are very similar. A control chart is the tool designed to make the time of manufacture groups become clear.

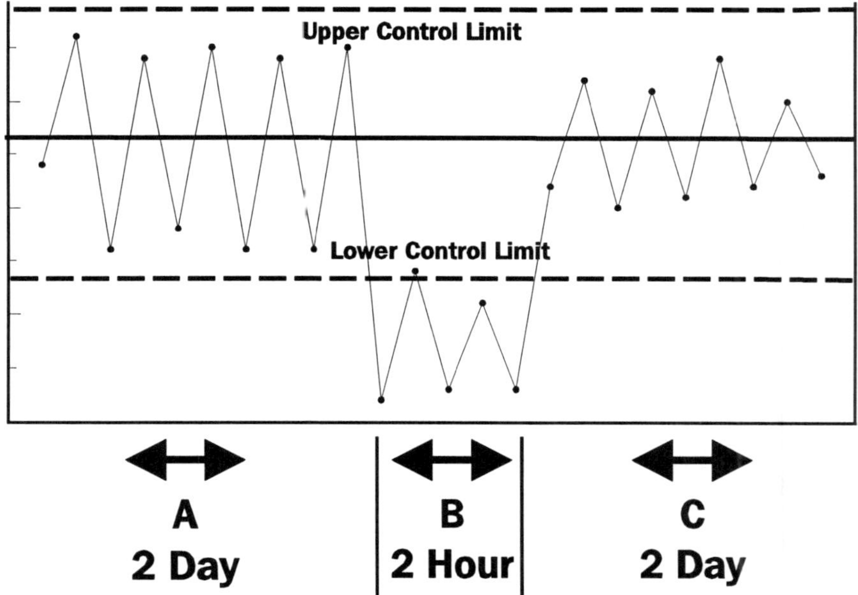

The three zones in this control chart must be understood and analyzed individually.

Take particular note of the different statistics from each time zone and then the difference for the lumping of the three zones.

Time Zone	Average	Standard Deviation	Distribution
A	50.00gm	1.0gm	Normal
B	46.55gm	1.0gm	Normal
C	50.00gm	1.0gm	Normal
A+B+C	47.33gm	1.0gm	Normal

In Chapter One we learned about normal distribution probabilities. These probabilities are computed from the average, the standard deviation, and computing the Z. The formula for Z is shown below:

$$Z = \left| \frac{X_{Ave} - X_i}{STD} \right|$$

The derived Z value is then used to compute the probability of an occurrence. For our candy problem, we have the following statistics and can compute their corresponding Z. Note how the different time zones have different percentage defective rates.

Time Zone	Average	Standard Deviation	Distribution	Z	Percent Defects
A	50.00gm	1.0gm	Normal	2.33	1.0%
B	46.55gm	1.0gm	Normal	5.00	.00%
C	50.00gm	1.0gm	Normal	1.55	6.0%
A+B+C	47.33gm	1.0gm	Normal	5.00	.00%

Time sequence grouping has the potential of presenting our process in very different ways. When the values are simply lumped together with no thought to proper group formation, disaster is on the horizon. Note how different the four possible analyses are.

A + B + C = 1%
A = 0%
B = 6%
C = 0%

Any decision based on improperly grouped (physical, associational, and time-series) data is a prescription for heading straight to Quality Hell.

Earlier we discussed the steps required for a capability study:

1) **Determine which data points to include**
2) Calculate standard deviation
3) Determine the specifications
4) Test for normality

Now the stage is set for the answer to the first question of *"What data points do we include in the capability study?"*

Determining Which Data Points to Include

First let's see how to decide which points to include in our capability study. This first step will use the knowledge that we built in our studies of control charts. For a machine, the process must be in a state of statistical control. Stated another way, the process must be consistent and predictable.

Our control chart labors will now be used to establish what time zones can be grouped together. Those time zones that are in a state of control can be grouped in one capability study.

IN CONTROL

Once the process is in statistical control (predictable), then a capability study can begin. The control charts are the tools to assure the process and product is in control. The capability study compares the process's abilities to the customer's requirements. This diagram shows that the results are consistent and in control and thus can be grouped together.

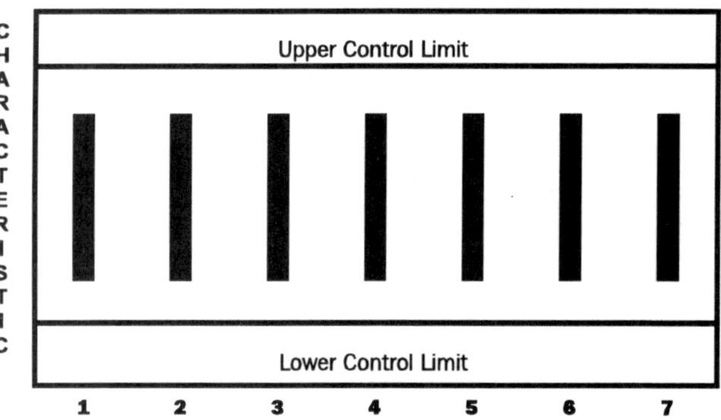

A Capability Study assumes that the process is in control (consistent and predictable). In Chapter Three, we studied the flaw of improper groups. The control chart is the means of assuring that a time-series group has been formed.

For the purposes of a capability study, use the data gathered when the process is in control. The in control time zones are shown on the control charts. The following chart shows a process that has shifted and whose values we should include. We will include all values for as long as the process is in control and, for this study, since the shift downward. This large number of data points (in control) increases the accuracy of our projections and thus minimizes the risk that we are making incorrect decisions.

This chart shows that the weight of the GooGoos produced from Machine 3 has changed. In Chapter Six, the SPC methodology of 1) investigate the cause, 2) alert management to the change, and 3) decide what to do were covered. The second step of this methodology alerts the strategic decision-makers to the change and triggers the need for a capability study.

From our candy machine, we can now identify the time zone to include in the capability study. From the control chart, the enclosed area is where the shift has occurred and is the area to include in the capability study.

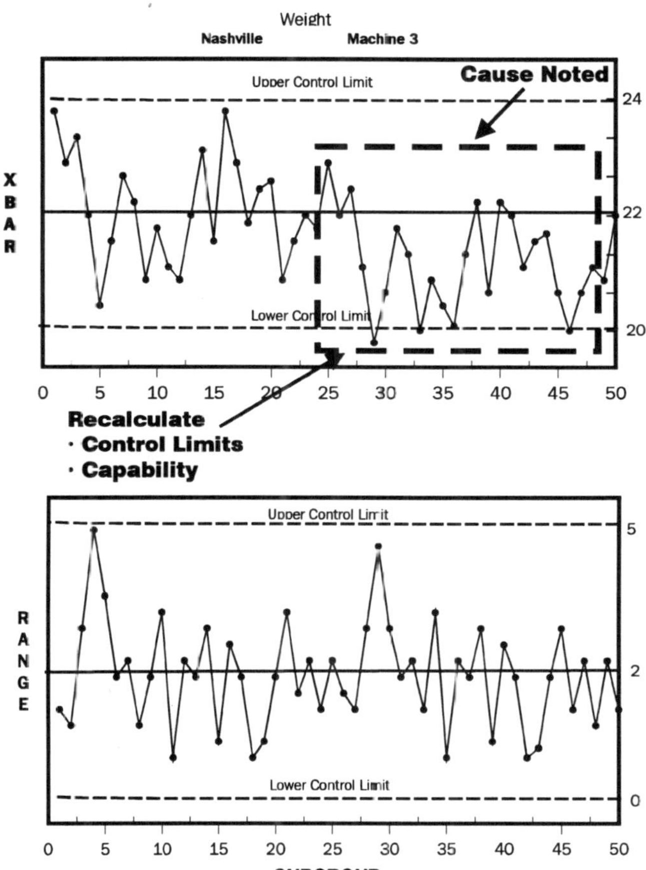

While the capability study is proceeding, the tactical forces are determining the cause of the shift.

In order to determine which values to include, we must know the process, when changes in process characteristics have taken place, (i.e., temperature, Ph, pressure, etc.), if the change is permanent, and if we will be retaining the change.

Now that we have a good appreciation for what data to include, let's return to the candy plant and start a study of the consistency of the chocolate machine's product.

A candy capability study of **Chocolate Consistency**

Let us now begin our capability study. The following readings are GooGoo chocolate consistency readings. Each reading is a test result from an individual GooGoo. Each row is the test results taken in a subgroup. Each subgroup is taken every 30 minutes.

TIME	OB1	OB2	OB3	OB4	OB5	X_BAR	Range
8:00	0.75655	0.76309	0.79555	0.85799	0.68696	0.77203	0.17103
8:30	0.73337	0.72810	0.83793	0.72416	0.68698	0.74211	0.15095
9:00	0.68130	0.68741	0.72701	0.64468	0.84080	0.71624	0.19612
9:30	0.60811	0.79303	0.90703	0.93568	0.75377	0.79952	0.32756
10:00	0.76692	0.62595	0.60721	0.69480	0.66086	0.67115	0.15971
10:30	0.73468	0.61227	0.66185	0.72843	0.55483	0.65841	0.17985
11:00	0.72128	0.75937	0.79861	0.81424	0.61291	0.74128	0.20133
11:30	0.76573	0.86283	0.66762	0.70669	0.61636	0.72385	0.24647
12:00	0.73689	0.80831	0.80004	0.69097	0.62378	0.73199	0.18453
12:30	0.67940	0.63179	0.64418	0.74433	0.70640	0.68122	0.11254
1:00	0.68170	0.77080	0.62278	0.57358	0.71586	0.67294	0.19722
1:30	0.84709	0.77807	0.59742	0.69553	0.81948	0.74752	0.24967
2:00	0.73915	0.80295	0.49851	0.81626	0.70454	0.71228	0.31775
2:30	0.87418	0.77712	0.92717	0.72820	0.76661	0.81466	0.19897
3:00	0.65501	0.79409	0.68381	0.76357	0.76406	0.73211	0.13908
3:30	0.77542	0.85875	0.54692	0.68720	0.71317	0.71629	0.31182
4:00	0.77416	0.72534	0.70543	0.64646	0.83704	0.73769	0.19058
4:30	0.60286	0.70500	0.68467	0.66985	0.67785	0.66805	0.10214
5:00	0.73883	0.67804	0.72443	0.68760	0.67617	0.70101	0.06267
5:30	0.78914	0.66123	0.62074	0.73161	0.86498	0.73354	0.24424
6:00	0.71017	0.77596	0.61749	0.65637	0.78831	0.70966	0.17082
6:30	0.70909	0.79797	0.72123	0.73636	0.77683	0.74830	0.08888
7:00	0.72207	0.57868	0.67210	0.79768	0.73411	0.70093	0.21900
7:30	0.67223	0.76367	0.65249	0.76464	0.67843	0.70629	0.11214
8:00	0.74534	0.77008	0.71271	0.62285	0.77221	0.72464	0.14936

The readings come from one chocolate machine with the chronological order of manufacture intact. Each time a reading is taken, five consecutive GooGoos are captured and tested. These five test points will be averaged to become our monitors on the control chart.

Control Chart Exercise

Using the data above, compute and prepare the data and control limits from the above chocolate consistency data. Once the computations are complete prepare the control charts and interpret them for a state of statistical control.

The following are the $\bar{X}$ Chart and the **R** Chart for the data points on the prior page. The control limits are computed for this chocolate machine. The $\bar{X}$ and R values are now posted on the control chart. The completed control chart now will allow us to determine which, if any, points can be included in the capability study.

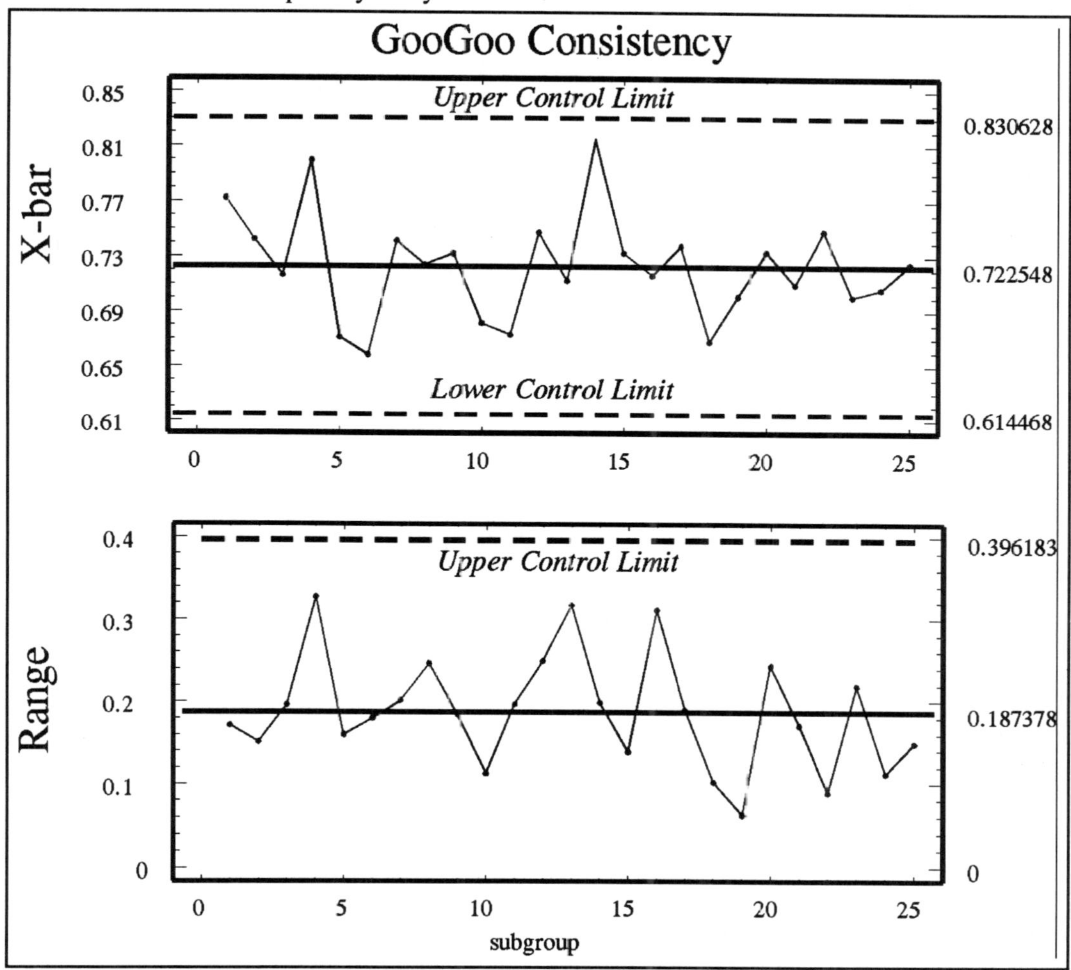

After reviewing the control charts, we can see no out of control points. No $\bar{X}$ or R values outside the control limits, no runs, and no trends are found. We can say that this ***process is in a state of statistical control***. This is a major accomplishment, but not to be confused with whether we are making chocolate with good consistency. We can now move forward to begin the capability study.

Let's return to the required preliminary steps for a capability study.

1) *Determine which data points to include*
2) **Calculate standard deviation**
3) Determine the specifications
4) Test for normality

Since we have now covered the first step and know what data points to include, the stage is set for the answer to the second question of, *"How do we calculate standard deviation in the capability study?"*

Calculating Standard Deviation

Analysis, decision-making, and goal setting tools are developed to support either strategic or tactical needs. These strategic or tactical tools fill a very precise goal and objective. The monitor for central tendency, the process average (strategic), is the same value as the control chart average (tactical). Remember there is a monitor of variability for strategic purposes and a second monitor for tactical purposes, but these values are different.

The function's strategic monitor of variability, standard deviation, is greater than the function's tactical monitor of variability. The strategic monitor is of individuals and is called **standard deviation**. The tactical monitor is of averages and is called **standard error**. The standard deviation is larger by a factor of the square root of the number of samples in the subgroup. These issues were fully covered in Chapter Two. The equation for the relationship of standard deviation and standard error is shown below:

$$\frac{STD_{Ind}}{\sqrt{n}} = STD_{Ave}$$

Our capability study is used for many strategic purposes from an initial assessment of the whether the process can perform its mission to determining the impact of a change that is detected by the control charts. First the clear understanding of the strategic use of this information is critical. Then we are ready to proceed with the mechanics of the capability study.

A capability study is used to assess the possibility of having an individual reading fall outside our specifications. Since we are dealing with individuals, strategic tools are required. Standard deviation is our strategic monitor of variability and is used in capability studies. The standard deviation of the individuals is used in comparing a process to a specification.

For the purposes of a capability study we must use the standard deviation of the individuals. We have two methods of calculating the standard deviation of the individuals. Note that both methods are estimates of the variability of the process and each will have a certain amount of error associated with it. First we will discuss the capability study's standard deviation formulas when used with a $\bar{X}$ and R Chart. Once these are completed, then we will show the differences for both the $\bar{X}$ and s Chart and individual charts. Now let's move into the $\bar{X}$ and R Chart variability monitors for the capability study.

$\bar{\mathbf{X}}$ Chart and the **R** Chart

$$Std_{Ind} = \sqrt{\frac{\sum_{i=1}^{N}(\bar{\bar{X}} - X_i)^2}{N - 1}} \quad \text{or} \quad Std_{Ind} = \frac{\bar{R}}{d_2}$$

As a general guideline, the sum of squares formula normally results in a larger value than the R estimate. The resulting smaller R estimator of variability will result in the percent defective being smaller for the R value estimate. Since *continuous improvement* is the normal objective, then I normally use the **Sum of Squares** as the estimator of standard deviation in capability studies. This larger standard deviation estimator yields the worst case or most percent defective estimate.

Sample Size	A_2	d_2	D_3	D_4
2	1.880	1.128		3.267
3	1.023	1.693		2.574
4	0.729	2.059		2.282
5	0.577	2.326		2.114
6	0.483	2.534		2.004
7	0.419	2.704	0.076	1.924
8	0.373	2.847	0.136	1.864
9	0.337	2.970	0.184	1.816
10	0.308	3.078	0.223	1.777
11	0.285	3.173	0.256	1.744
12	0.266	3.258	0.283	1.717

Specifications are the arbitrary requirements of our customers or management of a measurement attempting to assess the useability of our product. Specifications are referred to as tolerances, requirements, etc. We will discuss specifications in more depth a little later in the chapter. Simply having a product that meets the specifications is not a guarantee of a satisfied customer.

Specifications can be expressed as: Product is good only above a certain point, only below a certain point, or inside two measurement points. When we are calculating the capability (percent defective), we use the following test:

Specification	Capability Test
Product above a certain point	Less than (one tail)
Product below a certain point	Greater than (one tail)
Product within 2 points	Outside points (two tail)

Let see how we can use these variability monitors in our example.

Our candy consistency machine is using $\bar{X}$ and R charts to monitor the consistency of our chocolate candy clusters. Let's compute both monitors of the variability of chocolate consistency.

Our candy capability study

Let us now begin our capability study. The following readings are GooGoo chocolate consistency readings. Each reading is a test result from an individual GooGoo. Each row contains test results for a subgroup. Each subgroup is taken every 30 minutes.

TIME	OB1	OB2	OB3	OB4	OB5	X_BAR	Range
8:00	0.75655	0.76309	0.79555	0.85799	0.68696	0.77203	0.17103
8:30	0.73337	0.72810	0.83793	0.72416	0.68698	0.74211	0.15095
9:00	0.68130	0.68741	0.72701	0.64468	0.84080	0.71624	0.19612
9:30	0.60811	0.79303	0.90703	0.93568	0.75377	0.79952	0.32756
10:00	0.76692	0.62595	0.60721	0.69480	0.66086	0.67115	0.15971
10:30	0.73468	0.61227	0.66185	0.72843	0.55483	0.65841	0.17985
11:00	0.72128	0.75937	0.79861	0.81424	0.61291	0.74128	0.20133
11:30	0.76573	0.86283	0.66762	0.70669	0.61636	0.72385	0.24647
12:00	0.73689	0.80831	0.80004	0.69097	0.62378	0.73199	0.18453
12:30	0.67940	0.63179	0.64418	0.74433	0.70640	0.68122	0.11254
1:00	0.68170	0.77080	0.62278	0.57358	0.71586	0.67294	0.19722
1:30	0.84709	0.77807	0.59742	0.69553	0.81948	0.74752	0.24967
2:00	0.73915	0.80295	0.49851	0.81626	0.70454	0.71228	0.31775
2:30	0.87418	0.77712	0.92717	0.72820	0.76661	0.81466	0.19897
3:00	0.65501	0.79409	0.68381	0.76357	0.76406	0.73211	0.13908
3:30	0.77542	0.85875	0.54692	0.68720	0.71317	0.71629	0.31182
4:00	0.77416	0.72534	0.70543	0.64646	0.83704	0.73769	0.19058
4:30	0.60286	0.70500	0.68467	0.66985	0.67785	0.66805	0.10214
5:00	0.73883	0.67804	0.72443	0.68760	0.67617	0.70101	0.06267
5:30	0.78914	0.66123	0.62074	0.73161	0.86498	0.73354	0.24424
6:00	0.71017	0.77596	0.61749	0.65637	0.78831	0.70966	0.17082
6:30	0.70909	0.79797	0.72123	0.73636	0.77683	0.74830	0.08888
7:00	0.72207	0.57868	0.67210	0.79768	0.73411	0.70093	0.21900
7:30	0.67223	0.76367	0.65249	0.76464	0.67843	0.70629	0.11214
8:00	0.74534	0.77008	0.71271	0.62285	0.77221	0.72464	0.14936

Process Statistical Values

Average	0.72255
Standard Deviation of Individual	0.08047
Range Average	0.1873

Note the average of the individual is .7225, the standard deviation of the individuals (sum of squares method) is .08047, and the range average is .1873. Now let's take these values and compute the two alternative variability values for use in our capability study.

Sum of Squares Method

The standard deviation of the individuals using the sum of squares method is .08047. The formula and the results are shown below.

$$Std_{Ind} = \sqrt{\frac{\sum_{i=1}^{N} (\bar{\bar{X}} - X_i)^2}{N - 1}}$$

$$= .08047$$

We can also use the range average as the base for standard deviation.

Range Average Method

The range average of .1873 is used in our calculation of standard deviation of the individuals. The calculations and the results are shown below.

From the table on the right, the d_2 value of 2.326 is selected because the number of samples for the chocolate consistency is five. The d_2 constant is then inserted in the following equation. The standard deviation of the individuals using the range average method is .0806.

Sample Size	A_2	d_2	D_3	D_4
2	1.880	1.128		3.267
3	1.023	1.693		2.574
4	0.729	2.059		2.282
5	0.577	2.326		2.114
6	0.483	2.534		2.004
7	0.419	2.704	0.076	1.924
8	0.373	2.847	0.136	1.864

$$Std_{Ind} = \frac{\bar{R}}{d_2}$$

$$= \frac{.1873}{2.326}$$

$$= .0806$$

Notice the difference between the sum of squares method and range average is very small with the actual values of .0805 and .0806 respectively. Since the range average uses the part to part variation then this method should show the best or lowest value if there is group to group movement of the process. For our example, the two values are almost identical but this will not always be the case.

My rule of thumb is to always use the bigger of the two variability monitors. This will make the capability reflect the worst case scenario of having the most defects possible.

For our chocolate consistency problem, the standard deviation of 0.08047 will be used.

Other Control Chart Methods
The following tables and formula are used to compute the standard deviation of the individuals when those control charts are being used.

$\bar{\mathbf{X}}$ Chart and the s Chart

$$Std_{Ind} = \sqrt{\frac{\sum_{i=1}^{N} (\bar{\bar{X}} - X_i)^2}{N - 1}}$$

or

$$Std_{Ind} = \frac{\bar{s}}{c_4}$$

Subgroup Size	A_3	c_4	B_3	B_4
10	0.975	0.9727	0.284	1.716
11	0.927	0.9754	0.321	1.679
12	0.886	0.9776	0.354	1.646
13	0.850	0.9794	0.382	1.618
14	0.817	0.9810	0.406	1.594
15	0.789	0.9823	0.428	1.572
16	.763	.9835	.448	1.552
17	.739	.9845	.466	1.534
18	.718	.9854	.482	1.518
19	.698	.9862	.497	1.503
20	.680	.9869	.510	1.490
21	.663	.9876	.523	1.477
22	.647	.9882	.534	1.466
23	.633	.9887	.545	1.455
24	.619	.9892	.555	1.445
25	.606	.9896	.565	1.435

XBar & S Conversion Tables

Individual Charts

$$Std_{Ind} = \sqrt{\frac{\sum_{i=1}^{N} (\bar{\bar{X}} - X_i)^2}{N - 1}}$$

or

$$Std_{Ind} = \frac{\bar{R}}{d_2}$$

Subgroup Freq	E_2	d_2	D_3	D_4
2	2.660	1.128	-	3.267
3	1.772	1.693	-	2.574
4	1.457	2.059	-	2.282
5	1.290	2.326	-	2.114
6	1.184	2.534	-	2.004
7	1.109	2.704	0.076	1.924
8	1.054	2.847	0.136	1.864
9	1.010	2.970	0.184	1.816
10	0.975	3.078	0.223	1.777

Individual Chart Conversion Table

Again, the required preliminary steps for a capability study are:

1) *Determine which data points to include*
2) *Calculate standard deviation*
3) **Determine the Specifications**
4) Test for normality

Since we have now covered the first and second step, the stage is set for the answer to the third question, *"What are the specifications that will be used in the capability study?"*

Determining the Specifications Before a capability study can be prepared, a clear distinction between the customer's target and the specification must be established. This will allow us to separate what the customer wants from our abilities to achieve their desires.

Airline Example Since airline travel is part of my regular duties, let's use my last trip from Greenville, South Carolina, to Denver, Colorado, as an example. Imagine sitting on the plane all set to take off when the pilot says that we are going not to Denver International Airport (DIA) but instead to *somewhere around* Denver. He goes on to explain that enroute *he* will make a decision as to where to land. He continues that to add a little spice to the flight, we may land at the old airport (Stapleton), or possibly at the Boulder, Colorado airport, or possibly at the original destination of Denver International.

Thankfully, this did not really happen, but think about the consequences of such behavior. Passengers would miss their connecting flights, rental agencies would not know where to leave their cars, and other people meeting the passengers would be at a complete loss as to where to find them. A better way must be developed.

Just as a pilot must file a clear flight plan and stick to it, we, too, must know where we're headed before we begin a capability study. Two issues must be addressed. What does the customer really want? What can and will the customer tolerate today?

What Does the Customer Want?? When a customer purchases our product, he normally has a specific purpose or use in mind. Therefore one of our biggest challenges is to assure that we always keep this purpose in mind.

Target When we are traveling on an airline, our objective is to arrive in Denver, Colorado. When we go to the store for milk, we want one gallon of milk. When a half inch hole is drilled, we want the hole to be exactly a half inch.

For clarity of purpose, our target must be clearly identified. The target must be explicitly defined, and a value for the target must be established. This target value is a strategic decision. Some refer to the true target as the **nominal value**. The target simply defined is *the value that the customer expects should be perfect.*

Most customers do not expect perfection, but for our capability study to be effective, a clear understanding of what perfection would be is important. Remember this is called the **target**. If this is what the customer wants, then what are specifications?

What Will the Customer Tolerate Today?? Specifications are what the customer will tolerate today. Now let's expand on specifications.

> **Tolerance** 1. Leeway for variation from a standard. 2. The permissible deviation from a specified value of a structural dimension, often expressed as a percent.
>
> **Specification** 1. A detailed, exact statement of particulars, especially a statement prescribing materials, dimensions, and quality of work for something to be built, installed, or manufactured.

When the customer is describing his product, he communicates through tolerances and specifications. The customer has a specific item in mind. Specifications or tolerances are really the amount of *incompetence* that the customer will tolerate today.

When we are analyzing a product, we must have a clear, objective way of assessing its suitability. Generally this is done by specifications that may or may not be from the customer. We must continually evaluate how well our product is meeting the customer's intended use.

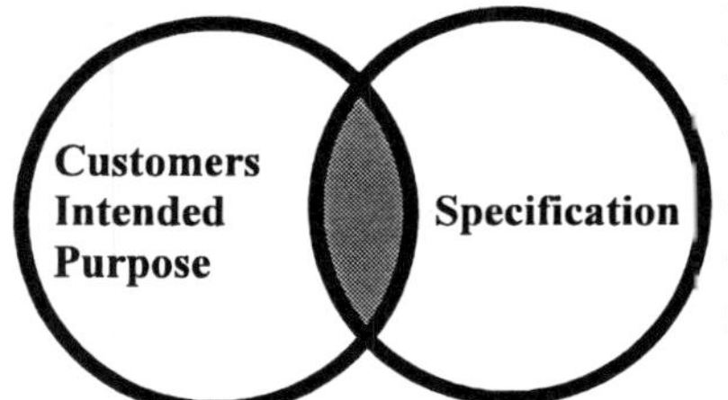

Product specifications are a simulated test of how the product will perform the customer's intended purpose. Care must be taken to assure a good overlap of intended purpose and the simulation. Frustrating consequences are caused between the customer and the supplier when we encounter the situation shown in the drawing on the left.

The drawing depicts the specifications no longer suiting the customer's intended purpose. Over time, we are not surprised to find our customer's wants and needs changing. Nor should we be surprised that the specification no longer represents the customer's needs. We must always be vigilant to assure that our specifications represent what he needs and wants.

If we use the degree of incompetence to explain specifications, we should not be surprised to see the customer's expectation fixed. We should also not be surprised to see the specifications continually changing. The tolerances or specifications are continually changing and getting tighter to drive toward the real expectations. The drawing to the right demonstrates how specifications will change over time until ultimately the customer receives exactly what he wants.

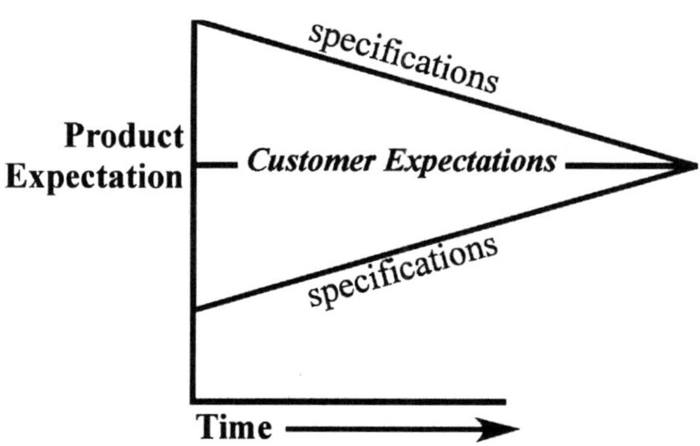

Both the target and the specifications must be defined by the strategic decision-makers. At no time should the tactician attempt to independently change these criteria. Any attempts for the tactician to take on the role of the strategist can lead to disaster.

Now let's visit the grocery store again to purchase a gallon of milk. Using the concept of specifications for a trip to the grocery store, rather than asking for a gallon of milk, we might ask for about 2 quarts to 1.5 gallons or something between buttermilk and water.

Properly established, the criteria should be a target of one gallon of milk. In the early 1900s when milk was obtained in the barn from Bessy the Cow, the level of incompetence was somewhere between two quarts and six quarts which was all Bessy had.

The amount of incompetency that was acceptable in 1900 was dictated by what Bessy could and would deliver.

Where do we use targets, tolerances, and specifications

Our communication must be clear to all associates, vendors, and customers as our requirements and needs. This communication is best accomplished through the use of stating the true target value. With this number stated the amount of incompetence is stated through the use of the plus and minus tolerances. Using our GooGoo candy example, the target weight is 50 grams. The amount of incompetence that we will tolerate is stated then with a tolerance of plus or minus 5 grams.

This now leads us to the specification that will be used in our capability study. We add the plus tolerance to the target to compute the upper specification limit. The minus tolerance is subtracted from the target to compute the lower specification limit.

The target and tolerances are used to communicate our requirements. The specification limits are used in the capability study.

Our Incompetency Drops as the Facility Changes In 1996 with milk in plastic and cardboard containers at the grocery store our level of incompetence has shrunk. The target is still one gallon, but now the specifications are almost exactly one gallon. Notice that over time, the incompetence index of the specifications will shrink. Also, this must be done without increasing the price.

Over time we must move toward the real specifications or target, or the consequences will be severe. For the grocery store example, we must close in on the one gallon or risk having to always provide more than the customer's expected amount. The option of giving more than a gallon in order to surely provide a gallon destroys our cost dimension. We are giving performance at the expense of cost. Remember our definition of quality: performance and cost are of equal merit.

What Are the Customer's Specification and Target for Our Exercise?? Our strategic decision-makers have set the chocolate consistency. For our candy machine, the target of the consistency is .7. Our customer specifications for the chocolate consistency is .5 on the low side and .9 on the high side. The target and the specifications must be used by all machines in the making of this candy.

Target	.7
Upper Specification	.9
Lower Specification	.5

The target and the specification limits are set by management to a target of .7 and an upper specification limit of .9 and lower limit of .5.

Once again, the required preliminary steps for a capability study are:

1) Determine which data points to include
2) Calculate standard deviation
3) Determine the Specifications
4) Test for normality

Now we must answer the last question, *"Is the data normally distributed for the capability study?"*

Testing for the Normal Distribution

Why is normality important? By having a normal distribution, a predefined set of probabilities can be used to compute the percent of values falling outside our upper and lower specifications. Not having a normal distribution is not disastrous but requires more work to determine the distribution for projection. Most times I find that when proper groups are formed and the process is in control, the normal does work well for projections. Many times the normal distribution comes naturally when the machine runs correctly and consistently.

How will we test for normality? Since the normal is not forced for capability studies and strategic decisions, a normality test must be used. There are several means of assessing the fit of our data to a normal distribution. One test, which was covered in Chapter One, is the use of a graph of the actual values presented as a bar chart or histogram compared to a perfect normal distribution for our data's average and standard deviation. This test requires some objective decisions and some margin for debate between two individuals.

A second option is to test for the distribution through quantitative means like chi-square or K-S goodness of fit tests. These quantitative means are outside the scope of this book.

The distribution of the individual data must be a normal distribution in order to do a classic capability study.

In order to test for normality, the following steps should be done.

STEPS

1. Determine and define the appropriate number of classes using the low value, the incremental value, and the high value.
    a. Determine the range of the X-axis
    b. Determine the number of classes. A good rule of thumb is approximately 7 to 10 classes. Too few or too many bins will cause the distribution to have an inappropriate appearance.
    c. Determine the range for entire data set.
    d. Determine the incremental value (class width) of the classes. Divide the range by the number of classes as selected from above.
    e. Determine the class boundary for each class.
2. Prepare the Histogram:

For a complete explanation of the preparation of the histogram, refer to Chapter One.

The graph below has been prepared so that we can determine if the chocolate consistency meets the criteria of the normal distribution. Passing this test will allow us to make predictions of the percentage defective to be expected from this machine in the future.

The criteria to test for normality was also covered in Chapter One. To review, most of the values are clustered around the average; as we move farther from the average, fewer values are observed. Finally the shape of the curve above the average is a mirror image of the curve below the average. Passing this criteria will allow us to use the predictions from the normal distribution table.

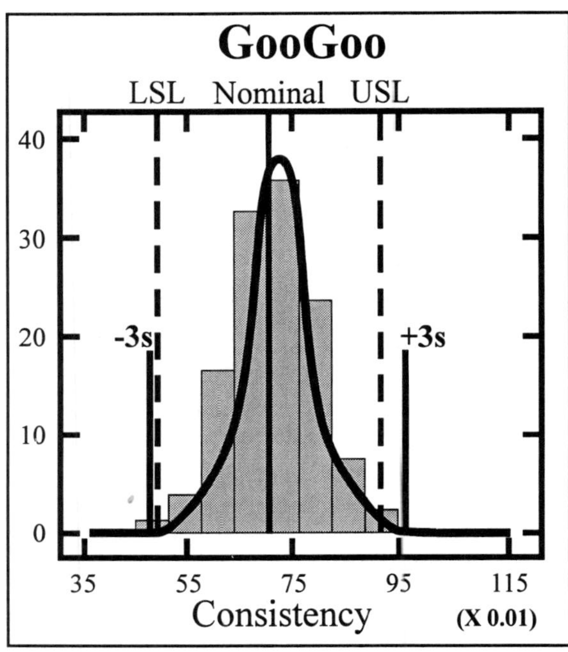

We can see that the chocolate consistency measurements of GooGoos from this machine are a normal distribution.

Begin analysis of a capability study only after determining normality. If the process tests non-normal, more advanced techniques must be applied to determine the proper fit of its distribution and probabilities.

The normal probabilities were covered in Chapter One. The calculation we learned there will be used to compute the Z values and, in turn, obtain the probability for projection. The following equation will now be used for our chocolate consistency example.

$$Z = \left| \frac{X_{Ave} - X_i}{STD} \right|$$

Recap of our preliminaries The preliminary work for the capability study is repeated below. Now our preliminaries for our candy exercise are shown below.

1) Determine which data points to include	The twenty-five sets of five candy clusters (125 individual clusters) will be used for our capability study.
2) Calculate standard deviation	The standard deviation of the 125 individual chocolate candy clusters is .0806. This value was computed using the sum of squares method.
3) Determine the Specifications	The following define the customer's criteria upper specification of .9, lower specification of .5, and a target of .7.
4) Test for normality	From our view of the histogram the data is declared to be a normal distribution.

Since we have now covered all the preliminaries, we are ready to begin the detail work of computing the capability study using the information that we have compiled.

Calculation Of Defects

The machine's capability can be calculated. We are ready then to compare the machine's knowns to the customer's knowns. These values are compared below:

Knowns

	About Our Process *versus*	**Customer**	
Average	.722548	Lower Specification	.5
Standard Deviation	.08047	Upper Specification	.9
Distribution	Normal	Target	.7

Our base formula from the normal table is shown below.

$$Z = \left| \frac{X_{Ave} - X_i}{STD} \right|$$

By rearranging the formula, we can calculate the distance from the central point to the spec limit where A is equal to the distance from the mean to the upper spec limit.

$$\text{Upper Spec Limit} = \bar{\bar{X}} + A$$

The value A is the distance away from the average to the upper specification limit.
If A is equal to ZS, then the formula would be:

$$\text{Upper Spec Limit} = \bar{\bar{X}} + Z\ Std_{Ind}$$

Solving the equation algebraically for Z will tell in units of S the distance away from the central point (i.e. 3 Standard Deviations)

$$Z(USL) = \frac{\bar{\bar{X}} - Upper\ Spec\ Limit}{Std_{Ind}}$$

This same formula is used for the lower specification limit. Since the Z is the absolute value of the distance (always a positive), then we can reverse the position of the specification limit with no impact on the results. This leaves us with the following equation for the lower specification limit.

$$Z(LSL) = \frac{\bar{\bar{X}} - Lower\ Spec\ Limit}{Std_{IND}}$$

Using our Normal Distribution table in appendix E, we can determine the percentage of failure above the upper spec and below the lower spec.

To shorten our expressions, the upper specification will be written X_{USL} and the lower specification limit is X_{LSL}.

Moving back to our candy machine problem, we can now compute the candy machine's capability.

Capability Analysis Example The following example provides the candy machine capability study for the upper specification limit. The information to be gleaned from this study is vital to improvement of the process. This provides the base camp of where we are to make our improvement steps.

What does the upper specification capability mean?? Capability is the percent of chocolate clusters that will likely be above the upper specification limit. Those clusters above the limit are then identified as defective; the consistency is not acceptable to our customers. This capability uses our consistent samples (in-control) to predict the expected number of occurrences for the future.

Upper Specification Capability

First our formula from the normal table derivation is brought forward.

$$Z_{USL} = \left| \frac{\bar{\bar{X}} - X_{USL}}{Std_{Ind}} \right|$$

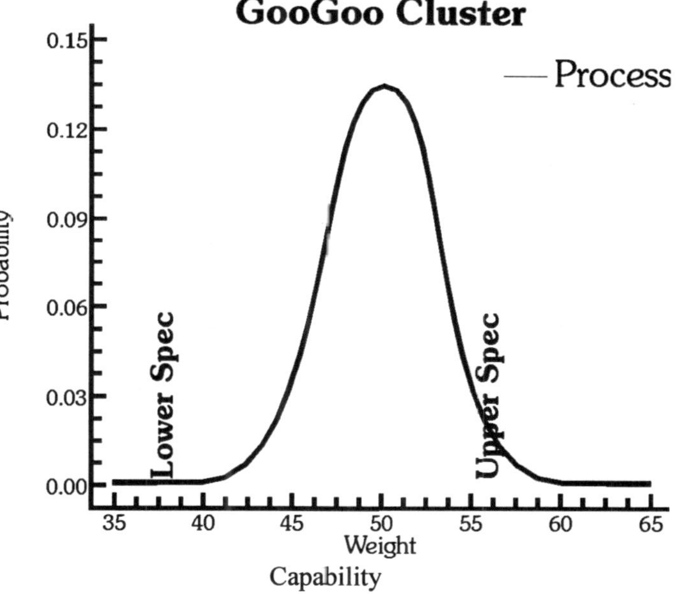

Next the process values of the average (.722548) and the standard deviation of the individuals (.08047) are inserted in the equation.

$$Z_{USL} = \left| \frac{.722548 - X_{USL}}{.08047} \right|$$

Now the customer's upper specification requirements (.9) are inserted in the equation.

$$Z_{USL} = \left| \frac{.722548 - .9}{.08047} \right|$$

Now the equation is solved to compute our Z value for the upper specification limit.

$$Z_{USL} = \left| \frac{.177452}{.08047} \right|$$

$$=2.2052$$

By using the Z table in the Appendix E, the percent out of specification can be projected. This percent defective above the upper specification limit is 1.37 percent.

$\% \text{ OUT}_{USL} = 1.37\%$

Lower Specification Capability

Now repeating the lower specification formula as done with the upper specification limit, our formula from the normal table derivation is brought forward.

$$Z_{USL} = \left| \frac{\bar{\bar{X}} - X_{USL}}{Std_{Ind}} \right|$$

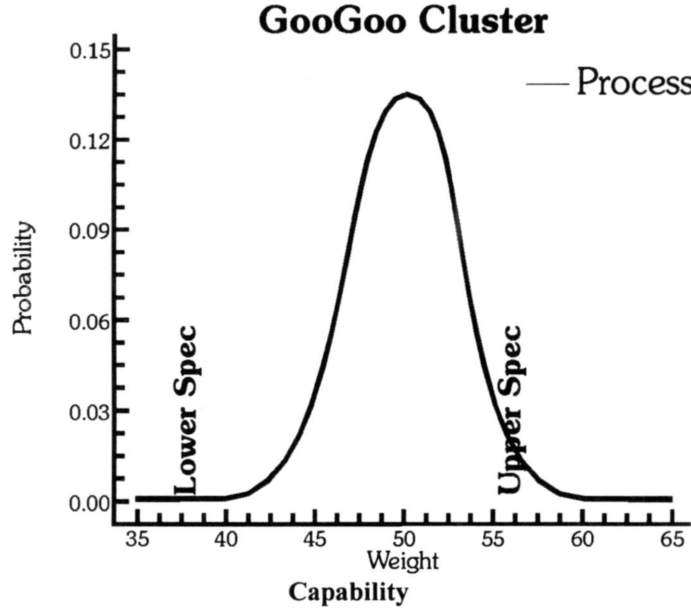

Capability

Next the process values of the average (.722548) and the standard deviation of the individuals (.08047) are inserted in the equation.

$$Z_{LSL} = \left| \frac{.722548 - X_{LSL}}{.08047} \right|$$

Now from the customer's requirement the lower specification (.5) is inserted in the equation.

$$Z_{LSL} = \left| \frac{.722548 - .5}{.08047} \right|$$

Now the equation is solved to compute our Z value for the upper specification limit.

$$Z_{LSL} = \left| \frac{.222548}{.08047} \right|$$
$$= 2.7656$$

By using the Z table in the Appendix E, the percent out of specification can be calculated.

$\% \text{ OUT}_{LSL} = .28\%$

The percentage defective below the lower specification limit and above the upper specification limit must be totaled to arrive at the total percentage defective projected for this machine.

TOTAL OUT OF SPECIFICATION

By adding the upper and lower specification percentages, the total percentage can be calculated. For this candy machine, the total percentage defective is 1.65 percent.

$$\begin{aligned} \% \text{ OUT}_{Total} &= \% \text{ OUT}_{UCL} + \% \text{ OUT}_{LSL} \\ &= 1.37\% + .28\% \\ &= 1.65\% \end{aligned}$$

Centering Distribution

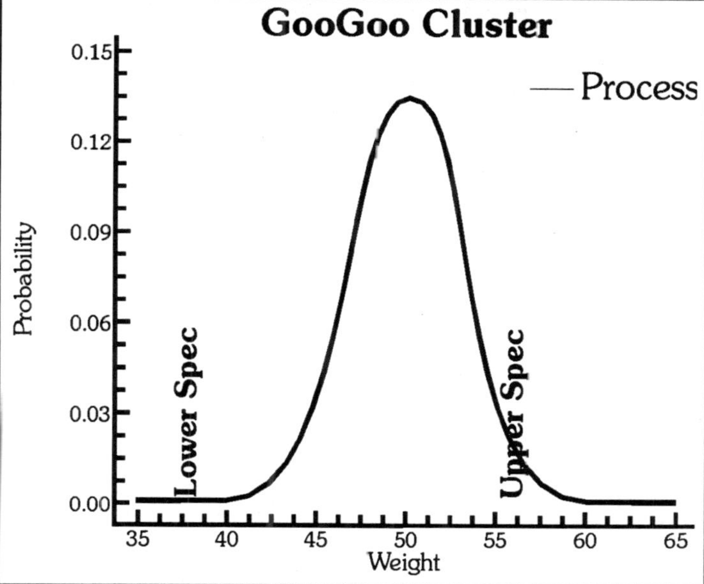

Capability

In Chapter One, the characteristics of a normal distribution were defined. The symmetrical nature of the normal provides an opportunity for improvement. Since the curve's slope is steep, if one of the specification limits is closer to the center than the other, the closer will have a larger percentage defective. Centering or keeping both specifications equidistant from the average can yield a reduced total percentage defective.

Since our candy machine has been identified to be a normal distribution, centering can reduce the percent total defective.

By changing the central point of the process in the direction of the greatest errors, we can reduce the overall percentage error. Not all processes can be adjusted, but when possible, saving and improvement are possible.

To calculate the optimum center point (average), the following calculation should be used. This center point is the spot to attempt to move the process's average if this is possible.

$$Center_{Opt} = \frac{X_{USL} + X_{LSL}}{2}$$

To calculate the percent out of spec, use the following equation and use normal tables.

$$Z = \frac{\bar{\bar{X}} - X_{USL}}{Std_{Ind}}$$

All specification limits are not of equal importance. In the case of drilling a hole an undersized hole can be reworked and corrected. A oversize hole cannot be reworked and creates scrap. Thus for some situations we may ***strategically*** choose to stay away from the most costly specification and not optimize our process.

Since both of the consistency specifications are of equal importance and cost, what would be the optimum average for our setting up of the process?

Candy Machine Centering Example

The following is an example of how to determine the optimum mean for lowest total out-of-specification.

$$\bar{\bar{X}} = \frac{X_{USL} + X_{LSL}}{2}$$

The customer upper (.9) and lower (.5) specifications are inserted in the equation.

$$= \frac{.9 + .5}{2}$$

$$\frac{1.4}{2}$$

This equation is then solved to yield a optimum process average of .7 chocolate consistency.

$$= .7$$

Now with this new optimum process average, our percent defective can be improved. Let's recompute our chocolate candy optimum average percentage defective and see the impact of this optimization.

$$Z_{OPT} = \left| \frac{\bar{\bar{X}}_{OPT} - X_{USL}}{Std_{Ind}} \right|$$

$$= \left| \frac{.7 - .9}{.08047} \right|$$

$$= 2.4853$$

By using the Z table in Appendix E, the % out of specification can be calculated as follows from our specification:

$$\text{\% OUT OF SPEC}_{OPTIMUM} = (.64\%)$$

Thus the total out of specification can be calculated as follows because the symmetrical characteristics of the normal distribution makes both specifications equally distant from the average and having the same percentage defective.

$$\text{\%OUT OF SPEC}_{TOTAL}\text{OPT} = .64\% + .64\%$$

$$= 1.28\%$$

Are These Optimization Savings Worthwhile? Now we can compare the original defective rate to the new optimum defective rate. The following equation allows us to compute the percent savings.

$$\%\ Savings = \%\ OUT_{TOTAL} - \%\ OUT_{TOTAL\ OPTIMUM}$$

The original percent defective is 1.65 percent and the optimum percent defective is 1.28 percent. These percent defectives can be inserted in our equation and the savings can be calculated.

$$\%\ Savings = \%\ OUT_{TOTAL} - \%\ OUT_{TOTAL\ OPTIMUM}$$

$$= 1.65\% - 1.28\%$$

$$= .37\%$$

When the equation is solved a .37 percent saving can be obtained.

$$\%\ Savings = .37\%$$

In a high volume manufacturing facility, these savings, when spread over time and for all machines, can lead to astronomical savings in the quality of performance, cost, and time.

What are Capability Indices??

Capability indices are tools to provide a monitor of capability in one package. These indices, when used correctly, can be an effective strategic tool.

All capability indices assume that the process and product are consistent and in-control. No capability index should ever be published on a machine that has not demonstrated it is in-control. Capability indices were intended for monitoring the capability of one particular machine.

The concept of capability indices

Many times management prefers to see an index that gives an indication of how well things are performing. The CP, CPK, and CPR are examples of indices that indicate how well our quality is performing. They are predicated on the assumption that the process is a normal distribution. (*Validate that assumption!*)

Since the normal distribution is the base for our capability indices, let's review the normal and its associated Z values. From the Normal Distribution we can find the Z values and percentage information:

Z	%
1	16%
2	2.3%
3	.1%
4	.003%

As the Z value goes up, the percentage goes down. For a machine, the lower the percentage the better the machine's quality performance. Stating this same idea with Z rather than the percentage reads: **the higher the Z, the better the machines performance.**

As the Z value goes down, the percentage goes up. For a machine, the higher the percentage the worse the machine's quality performance. Stating this same idea with Z rather than the percentage reads: **the lower the Z, the worse the machine's performance.**

Most of our capability indices want to reflect the worst case performance for the machine under study. Our first indices will be an understanding of a index called CPK.

How does CPK work??
The **CPK Index** is the indicator of the worst case specification and is calculated according to percentage defective. This ratio is the Z for the worst case percentage defective compared to a defined value of 3. This can be converted into the equation that follows:

$$CPK = \frac{Minimum\ (\ Z_{LSL}\ ,\ Z_{USL}\ )}{3}$$

This equation can then be simplified to reflect the Z_{Min} that is the worst case percentage defective of the two specification limits. This equation is shown below:

$$CPK = \frac{Z_{MIN}}{3}$$

A CPK value of one (1) indicates the worst specification equates to the Z score of 3.0. ***The bigger the CPK the better the process is functioning when performance is the only concern.***

Candy Machine CPK Index Example

The machine's capability was calculated and can compute the machine's CPK from our initial capability study. The initial capability used the following knowns for the process and the customer requirements. These values for comparison are shown below:

Knowns

About Our Process	*versus*	Customer	
Average	.722548	Lower Specification	.5
Standard Deviation	.08047	Upper Specification	.9
Distribution	Normal	Target	.7

From the capability example, we have been using, let's analyze the capability of this process. The low specification limit has a capability of .28 percent defective and a corresponding Z of 2.7656. The upper specification limit has a capability of 1.36 percent defective with a corresponding Z of 2.2052. The table below summarizes the capability data from which the CPK will be computed:

	USL	LSL
Z	2.2052	2.7656
%	1.36%	.28%

Note, our efforts are to show the worst case percentage defective for this machine. The worst case is the higher percentage or the lower Z value. The minimum Z value or the highest percentage out of specification is the USL percentage of 1.36% or Z = 2.2052.

To calculate the CPK for our example, we use the following equation:

$$CPK = \frac{Minimum(Z_{USL}, Z_{LSL})}{3}$$

The upper specification Z of 2.2052 and the lower specification Z of 2.7656 are inserted in the equation.

$$CPK = \frac{Minimum(2.2052,2.7656)}{3}$$

Now the equation is solved to yield the following results:

$$\text{CPK} = \frac{2.2052}{3}$$

$$= .7351$$

The CPK for this machine is .7351. **The bigger the CPK the better the performance of the machine. When all other things are equal for two machines, the machine with the larger CPK is the better performing machine.**

What are some other indices?

The **CP Index** is an indicator of both the upper and lower specification limit related to a six (6) standard deviation. A one (1) indicates that the specification and the control limits are the same.

$$CP = \frac{Z_{USL} + Z_{LSL}}{6}$$

The **CR Index** is the inverse of the **CP Index.**

$$CR = \frac{6}{Z_{USL} + Z_{LSL}}$$

Never use a capability study without a clear understanding of all the components of statistical process control (statistical monitors, control charts, and a complete capability study). The simplistic use of one monitor has lead many to the depths of Quality Hell.

Management Indicators
These capability indices require some knowledge of statistics. A preferable means of noting the capability follows. First we must revisit our complete definition of quality which includes performance, cost, and time. Then we will build a index around the complete definition.

From our prior tactical and strategic steps, we have a set of knowns. The control chart has validated that the process is consistent and predictable. The capability study has projected the percent product out of specification.

Any manufacturing facility will know the volume of parts produced per day and the unit cost.

Now all these pieces of information can be compiled to compute one cost of quality measure.

Management Quality Opportunity Cost Formula
The quality opportunity cost is derived by first computing the percentage of defects, then the percentage of defects are multiplied by the total units produced per day. The total defects/day are then multiplied by the cost per unit. The result is the cost of defects per day.

Quality Opportunity Cost = Percent Defect * Units Produced * Cost/unit

We can use the above variables to calculate the total cost of our quality. With this information, management can make strategic decisions as to where to use our limited resources for the most good.

The type of decisions that can be made include the following:

1. Return on Investment
2. Capital Investment
3. Preventive Maintenance
4. Areas to Direct Our Improvement Resources

MANAGEMENT INDEX EXAMPLE
The following are the results of our prior steps in the SPC process:

The candy machine is running in a state of statistical control or consistently.
The percentage defective is 1.65 percent out of 100,000 units produced on this machine every day.
The cost of every chocolate candy cluster is $.20 per unit.

Knowns Restated

Process is in Control

% Out Spec	1.65%
Production per day	100000 units
$/Unit	$.20

Production Cost
The product cost for this one candy machine is the unit cost multiplied by the units produced. For our example, the unit cost is $.20 and the units produced are 100,000 per day. This yields a cost of $20,000 per day for this machine.

Management Index Calculations for our candy example

(% Out Spec) (Production) ($/Unit)

(1.65%) (10000 per day) ($.20 per unit)

$330

Opportunity Cost Interpretation
The $330 opportunity cost must be weighed against the total cost of $20,000 to keep the opportunity cost in perspective. Now we have a grasp of the magnitude of the potential return from any improvement investments.

These values can be used for strategic decisions as to the potential viability of any project. As an example, if our engineers tell us to eliminate the $330, an enhancement costing $2,000,000 might be strategically rejected. A $10 minor adjustment saving half the cost might be accepted.

Now our strategic decision-makers can function with clean, sound information and intelligence.

Capability Exercise

The GooGoo components' cost per gram is $.005. With this information, we can begin to analyze our strategic options. In our original proposal, the cost is based on a GooGoo that weighs 45 grams. By multiplying the weight of 45 grams times a component cost of $.005/gram, *the anticipated cost of a GooGoo is $.225.* For an hour's run, the cost of 3000 GooGoos would be 3000 GooGoos times $.225/GooGoo equaling $675.

The GooGoo facility is being run correctly and consistently as demonstrated by a control chart.

The following four situations are provided for us to investigate. Imagine that this is your candy company. Compute the capability of each and make a recommendation as to the best situation and why.

Situations	1	2	3	4
Is the Process In Control?	Yes	Yes	Yes	Yes
Average Weight	50gm	52gm	47gm	50gm
Standard Deviation of Weight	1.0gm	1.0gm	1.0gm	2.0gm
Distribution of Weight?	Normal	Normal	Normal	Normal

Chapter Summary

The capability study is the strategic arm of SPC. This tool allows the strategic decision-maker to understand whether the facility is able to produce the customer's product. Once the capability is established for a machine, any changes that are detected by the tactical control chart can be assessed for impact on the customer.

The capability study must be used only in its total context of statistical process control where control is handled by the tactical tool of the control chart. Before we do a capability study, the process must be in a state of statistical control.

The Process Control Method The overall objective of Statistical Process Control (SPC) methods is to prevent product or performance defects.

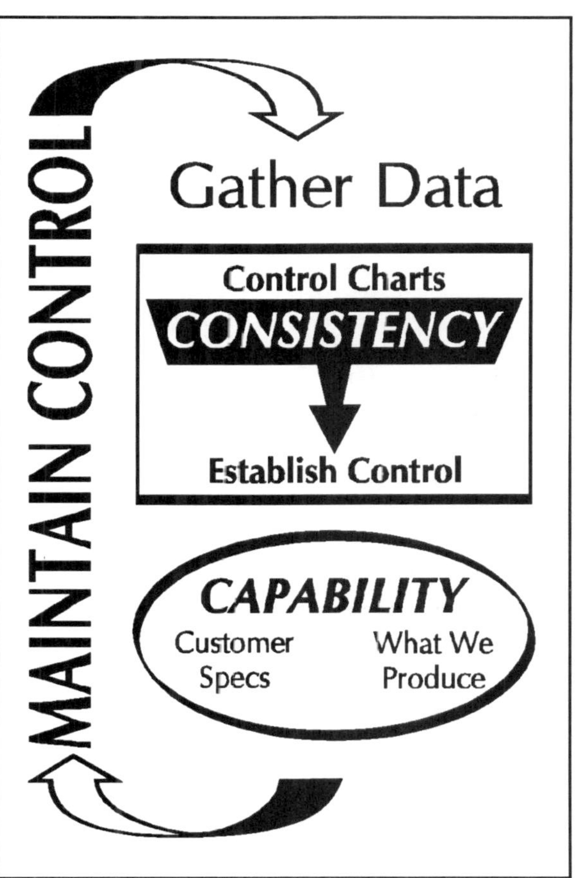

How many defects should we tolerate? The aim of process control is to prevent defective product. That requires us to predict the percentage of defects from the controlled process. In turn, this requires that we know about the distribution of values from the process itself.

However, our data at this point is the averages of samples drawn from the process for the tactical purpose of running the facility - not the process values themselves. We can use the same sample data that appears in our control chart to estimate the location and spread of the process distribution, and thus the proportion of product that falls outside of the customer's specifications. To this end, we must analyze the values with strategic tools called prediction limits which use the individual readings.

The capability study uses the strategic set

of calculations called prediction limits. The following summarizes the two complete sets of books available, and which decision-making process will use the appropriate analysis set of books. Strategic decisions use prediction limits while tactical decisions use confidence limits.

	Tactical	Strategic
	Operation ↓ Detect Change ↓ Averages	Management/Customers/Sales ↓ Assess Capability ↓ Individuals
Central Tendency	**Ave (Ave)** =	**Ave (Ind)**
Variability	$\mathbf{STD_{Ave}}$	$\mathbf{STD_{Ind}}$
	$\mathbf{STD_{Ave}} = \frac{\mathbf{STD_{Ind}}}{\sqrt{\mathbf{n}}}$ Where n = subgroup size	
Distribution	Central Limit Theorem Normalizes Data	Histogram
	Confidence Intervals $X_i = X_{Ave} \pm z * \frac{Std_{Ind}}{\sqrt{n}}$	Prediction Intervals $X_i = X_{Ave} \pm z * Std_{Ind}$

For this chapter on capability studies, the prediction limits and the standard deviation of the individuals are used.

A capability study provides management with a tool to assess the ability of a machine to make what the customer wants. This same tool is used to assess the impact all changes have on the customer.

Concluding Thoughts: Tactics and Strategy Working Together

Performance

Quality

Time Cost

Our definition of quality is extremely broad and encompassing. According to that definition, quality is comprised of three components: performance, cost, and time. Our research, analysis, and decisions must support our goal of delighting our customer (Performance), while reducing our cost (Cost), and providing our product to our customers the moment they want it (Time). We must view this goal as a never-ending journey that continues to improve all three components because no one component is more important than the others. All three must be viewed with equal importance and balance. This balance requires the Business Genius of the Manager of the Future.

The improvement model and statistical process control work together to drive improvement. The model below that was covered in the preface, and SPC are the tools to improve our operation.

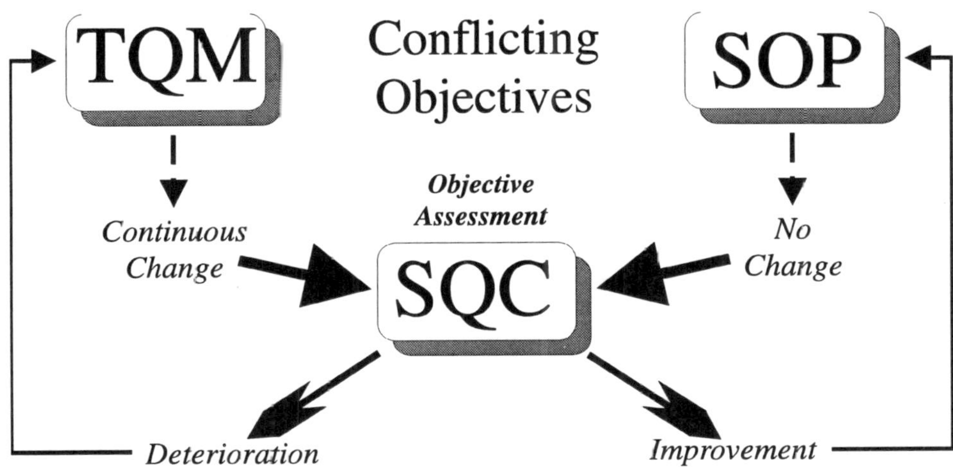

Improvement Model

Do you remember Barney Fife in the classic television program *The Andy Griffith Show*? He always told Sheriff Taylor that they had to "nip it, nip it, nip it in the bud" whenever they encountered a problem. SPC and the improvement model are like that; we must nip it in the bud! Find every small problem and correct it before it becomes a big problem. Find every improvement and put it into the standard operating procedure.

Obstacles to Quality The statistical process control effort to support our quality is critical. As a final set of thoughts, the following are obstacles that will impede your efforts at improvement if they are not addressed directly.

1. Use of averages as a stand alone indicator of production, cost, quality, etc.

2. Lack of a clear understanding of consistency and the term "In-Control"

3. Machines, departments, etc. lumped together for analysis because it's expedient. Using this information to develop production, cost, or quality values without regard to mixing of unlike products or processes.

4. Loss of the traceability of our product. Failure to know when the goods were processed makes any time analysis impossible.

5. Focus on product characteristics only. Failure to monitor and control process characteristics. Failure to understand what is a *process* and how it works.

6. Unidentified and undocumented assumptions and speculation.

7. Use of opinions as opposed to the correct use of facts.

8. Failure to move SPC to the shop floor, because we assume the operators cannot or will not understand SPC.

9. Failure to understand division of labor. Who should be accountable for what is essential to producing a quality product. (Management **and** operations)

10. Customer specification on a Control Chart. This is the most common mistake made in the use of SPC. Trying to infer good or bad product from a control chart is impossible.

11. Failure to have a metric blueprint.

12. No willingness to make changes.

13. No plan for implementing SPC and the improvement effort.

14. The strategic decision-maker failing to provide adequate and proper resources to assure the successful implementation of the plan.

15. Failure to execute the plan.

Inadequate attention to these obstacles will doom you to remain in that dismal, fiery, awful place called Quality Hell.

Control charts and capability studies are the two tools of SPC. They can and will when used correctly form a team that will keep us on the straight and narrow path to the improvement mountain top.

Our strategic tool of capability studies will allow the strategic decision-makers to have the information to make effective decisions.

Our tactical tool of control charts will allow us to understand that this machine is running consistently.

The SPC team

Both tactical and strategic tools must work together as a team to allow our improvement model to work effectively. Neither tool is more or less important than the other but both must work together as a team just as the baseball pitcher and catcher form a team. The pitcher cannot be successful without a good, effective catcher. Capability studies cannot be effective without good, effective control charts.

Process Improvement is an Iterative Procedure

Recalculating Control Limits: Once the process has been determined to be "In Control", the control limits should be locked. When a process change has been determined and is desired or permanent, recalculation of the control limits is required. Control Limit Basis is determined by the subgroups to include in the calculation.

Get Control
- Compute trial control limits
- Review and analyze charts
- Eliminate special causes
- Collect more data
- Repeat until stable

Define Current Capability
- Assure that the process is stable
- Access capability - Know **Process** Standard Deviation
- Compare to customer specifications using a histogram
- Determine need for management action

Once the SPC methodology is in place, a continuous surveillance of the control charts is required. The following is reflective of an alarm having been detected and a cause having been established. Once the cause is determined and it is ongoing, the control limit and companion capability study must be recalculated.

When an alarm is detected, the following steps are required:
- **Investigate** to determine the reason (cause) for the alarm. This investigation is a tactical operation for the operational staff.
- **Determine Capability.** Prepare a capability study to determine the impact (deterioration or improvement) on the process. The assessment of the changes impact is a strategic effort accomplished by the management staff.
- **Take Action.** Develop and implement an action plan to deal with the cause. *Remember we are acting on the cause not the alarm.* The action is a joint responsibility of the tactical and strategic team. Both management and operations must work in harmony to develop the best plan of action.

Please take special note that an out-of-control condition does not imply bad product. Out-of-control is intended to draw our attention so that we can investigate for the cause. We may find the alarm has highlighted an improvement.

Recalculating Capability And Control Limits

The control limits and capability studies should only use the data that is reflective of the current status. This can be seen in the area noted as *Cause Noted* below:

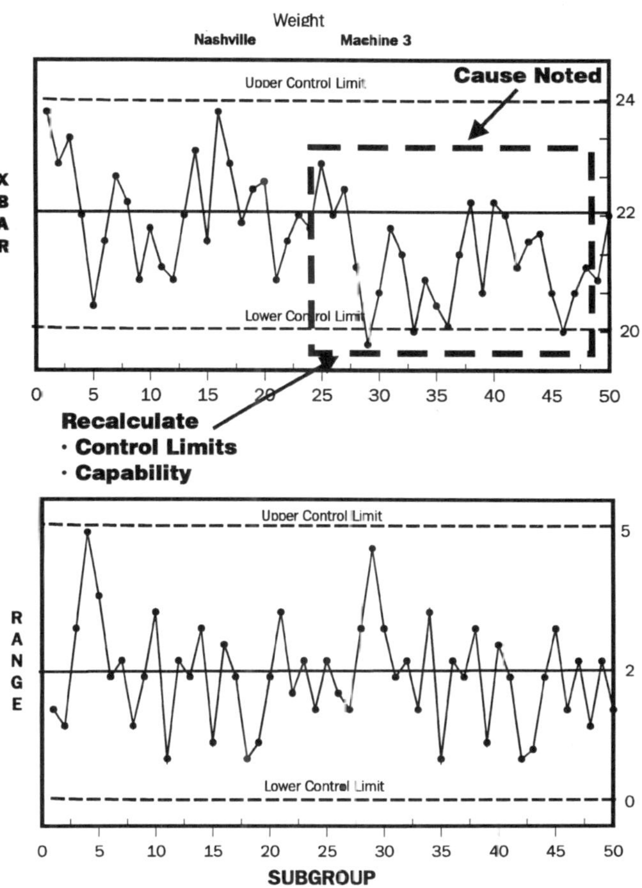

If this current status is based on a relatively small number of data points, the control limits and capability may be made temporary. When more data is gathered, recalibration of the limits can be calculated and then locked into place.

We have discussed at length the monitoring, control, and capability of the product, but this same methodology applies to all product and process characteristics. The timeliness of the data, the analysis of the data, and the sound logic of decisions is crucial to the success of product disposition.

Once the process has been determined to be in control, the control limits should be locked. When a process change has been determined and is desired or permanent, recalculation of the control limits is required. The control chart, control limits, and capability study must all be based on the same time frame. This is required to allow the control chart methodology to function as an alarm system. This alarm system will tell us when the process is either improving or deteriorating.

What is the Team for Decision Making?

Tactical execution is required for strategic decisions to be effective. We have already discussed the consequences of incorrectly running the facility. In such cases, no strategy will work. Our strategic decision-makers must count on our tactical decision-makers to do their jobs, for only then can the strategic decisions be effective. Many strategic options are available, but they will work only if the operation can run smoothly, or in a state of control.

To repeat, management can make decisions only when operations is correctly and consistently running the facility. When this is the case, many options are available to management. When the plant is run incorrectly, we have no way of making a strategic decision that is based on sound logic.

When strategic and tactical personnel both do their job, they form a team. Operations is in deep trouble when management does not do its decision-making job of providing the facility. If the facility (available means) is not capable, we are doomed to having dissatisfied customers and stockholders. If operations does not perform its role of running the facility (available means) correctly and consistently, then how can management assess the impact to the customer?

When operations (tactician) and management (strategist) execute their jobs brilliantly, we can truly improve our quality. We can impact all three components of quality: performance, cost, and timeliness. Let's see an example of how tacticians and strategists should perform.

Now let's return to our candy facility. We are concerned that our customers are always satisfied, but we must take care that our company remains profitable. We must investigate a machine for determining our optimum running criteria which will please both our customers and our stockholders.

Let's do a problem to assess several situations for our candy machines.

The GooGoo components' cost per gram is $.005. With this information, we can begin to do analysis of our strategic options. In our original proposal, the cost is based on a GooGoo that weighs 45 grams. By multiplying the weight of 45 grams times a component cost of $.005/gram, *the anticipated cost of a GooGoo is $.225.* For an hour's run, the cost of 3000 GooGoos would be 3000 GooGoos times $.225/GooGoo equaling $675.

The GooGoo facility is being run correctly and consistently as demonstrated by a control chart.

Strategic Option 1

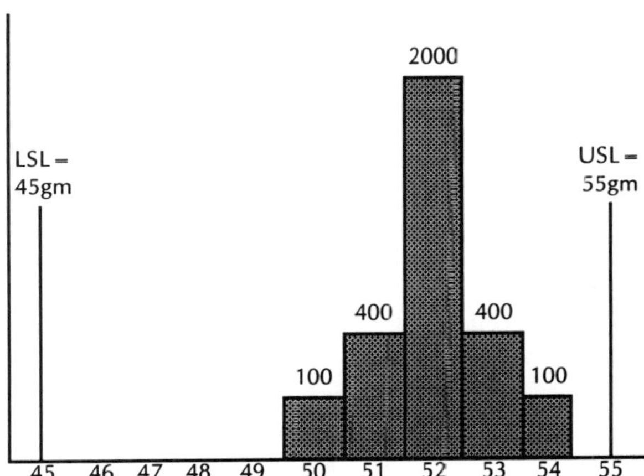

Buying a Delighted Customer
A decision can be made to always give the customer *heavy* GooGoos. This choice guarantees happy customers but our costs are high. Our package states that no GooGoo shall weigh less than 45 grams. Since we are selling, advertising, and promoting 45 gram GooGoos, our customers will give us no additional credit for the additional GooGoo weight.

In this option, we will use the same cost base for the component cost of $.005 per GooGoo gram. By multiplying the weight of the GooGoos by a component cost per gram by the number of GooGoos, we can compute the cost of an hour.

Weight in grams	Cost per Gram	Quantity Produced	Total Cost
50	.005	100	$ 25
51	.005	400	$102
52	.005	2000	$520
53	.005	400	$106
54	.005	100	$ 27
	Total Cost		**$780**

Comparing the original estimate of $675 to our current cost of $780, we are spending an additional $115 per 3000 GooGoo or 3.8 cents per GooGoo to keep our customers happy.

In our second option, the GooGoo facility is still being run correctly and consistently as demonstrated by a control chart.

Strategic Option 2

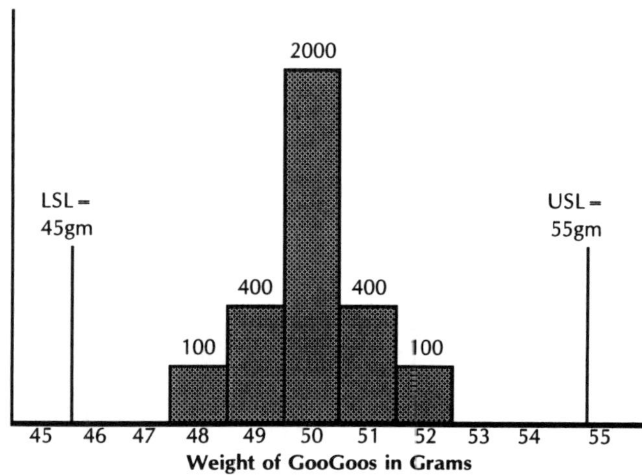

Conservative Strategy
This strategic option is the mid range. We are in the middle of customer specifications. Our customers will still be pleased and our cost is in the mid range.

In this conservative option, the cost is still based on our component cost of $.005 per GooGoo gram. By multiplying the weight of the GooGoos by a component cost per gram by the number of GooGoos, we can compute the cost of an hour.

Weight in grams	Cost per Gram	Quantity Produced	Total Cost
48	.005	100	$ 24
49	.005	400	$ 98
50	.005	2000	$500
51	.005	400	$102
52	.005	100	$ 26
	Total Cost		**$750**

Comparing the original estimate of $675 to our current cost of $750, we are spending an additional $75 per 3000 GooGoo or 2.5 cents per GooGoo to keep our customers happy.

Again, the GooGoo facility is being run correctly and consistently as demonstrated by a control chart.

Strategic Option 3

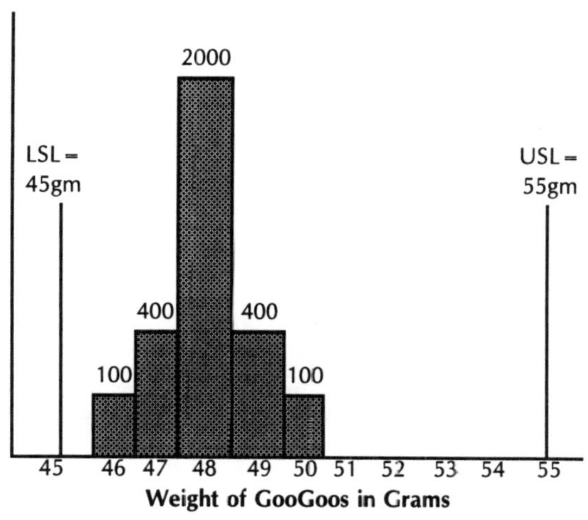

Satisfied Customer and Lower Cost

This strategic option will keep our customers happy while keeping our cost at a minimum. This option would make our shareholders happier because cost is at a minimum and our customer is getting exactly what the bargain required.

In the final option of satisfied customer and low cost, the cost is still based on our same component cost of $.005 per GooGoo gram. By multiplying the weight of the GooGoos by a component cost per gram by the number of GooGoos, we can compute the cost of an hour.

Weight in grams	Cost per Gram	Quantity Produced	Total Cost
46	.005	100	$ 23
47	.005	400	$ 94
48	.005	2000	$480
49	.005	400	$ 98
50	.005	100	$ 25
		Total Cost	**$720**

Comparing the original estimate of $675 to our current cost of $720, we are spending an additional $45 per 3000 GooGoo or 1.5 cents per GooGoo to keep our customers happy.

Summary of the Strategic Options

Let's compare our three strategic options. The cost of the options is shown in the following table.

Strategic Option	Total Cost of Each Option
1 **Buying Delighted Customers**	$780
2 **Conservative Strategy**	$750
3 **Satisfied Customer and Lower Cost**	$720

No one strategy is correct for every situation. All of the strategies above have equal merit, and for each different situation, we will select a different best option. Each strategy must be viewed in the total relationship of customer, cost, profit, future potential market, product, and liability. No strategic decision can be made unless operations has done its job.

As can easily be seen, management is dependent on operations doing its job of running the facility correctly and consistently. Similarly, operations is clearly dependent on management to provide *the correct facility* for the particular job. When the facility runs correctly and consistently, an option can be selected. For our candy company, having **Satisfied Customers and Lower Cost** at a cost of $720 per 3000 GooGoos would be our selection. This option allows us to keep our customers happy and save $.02 per GooGoo compared to our highest option. **These savings are only possible if the process is run correctly and consistently.**

SPC teamwork

The SPC team of control charts and capability studies, when functioning together, will allow us to hit the bulls eye of improvement, customer satisfaction, reduced cost, and timely deliveries.

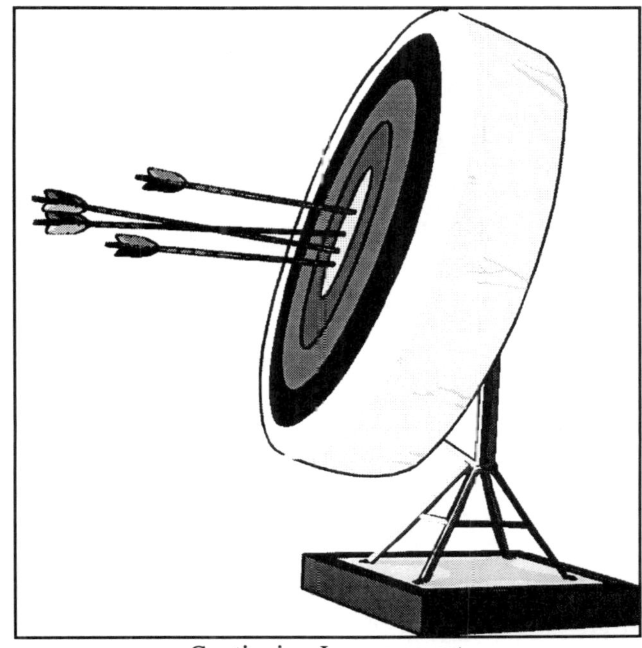

Continuing Improvement
...Through Control and Capability

Remember that the control chart limits are an extension of the capability study. Thus as long as no alarms are detected on the control limits, no change (neither improving or deteriorating) has been detected. Thus when a change is detected that leads to a long term deterioration, then the control limits must be recalibrated to reflect this change to keep the tactical and the strategic efforts in sync.

Understanding Capability by Implementing SPC

Our SPC team of control charts and capability studies form our information team for climbing the improvement mountain top. The tactical tool of control charts give us a clear picture that a machine is running correctly and consistently. Then the capability study provides the strategic decision makers with the assessment that the machine can make what the customer wants.

The climbing of a mountain requires a sturdy and strong foundation. This mountain-climbing foundation is the base camp where all ascensions are launched. Our base camp is the knowledge that SPC provides.

With this information in hand, our decision-makers are ready to make improvements. This base encampment of our current knowledge base contains the correct and consistent running of each machine and the capability of each machine to make our product.

What Information Can We Pass Along?
Our information is our machine's process and our product knowledge base. This knowledge base is the tactical knowledge that the machine is running correctly and consistently. Supplementing the tactical knowledge is the strategic knowledge of the capability of this machine. We must use the capability of each machine to then develop a plan for dealing with the results.

How will we use this Information?
First we can use our base camp of knowledge to launch all our ascensions to the Improvement Mountain Summit. With clear knowledge in hand we can make our improvement plans. Each improvement plan is like a mountain climbing ascension to the summit of a mountain. The improvement project plans must be based on clear facts. Each ascension is a project plan for improvement.

One ascension is a plan for improvement by making informed future expansions and the purchase of new equipment. Any new equipment or expansion that is built without the base camp will be risky. The missing base camp of knowing how to correctly and consistently run the existing equipment is a bad operational habit. This habit used on the existing old equipment can cause the old bad habits to jeopardize the new equipment. The bad habits are hard to break and when our tactical work force uses the old habits of no process knowledge on the new or expanded equipment bad results are sure.

An ascension of dealing with noncapable machines must be based on the assurance that the machine's capability is based on correctly and consistently run equipment. With our tactical knowledge in place we will know which machines to retool and what is in need of repair. We will also know which machines need to be replaced.

Now we will know the importance and magnitude of each improvement. This information will tell us how and where to make major and minor improvements. With the importance known we can wisely make our strategic decisions as to where to spend our resources of time, money, and people. Strategic priorities can be placed on all our improvement projects to give us the biggest bang for our limited resources. Now we are spending our resources for the very best value.

Finally we can use our strategic and tactical information to achieve the original objective of quality. This objective will be achievable since we are sound in the knowledge that each machine is run as correctly and consistently as possible. With our tactical issues clear and validated then the machine's capability is known and accurate. Now our original objective of quality is in sight - it is the summit of the Improvement Mountain.

Quality is comprised of the three components: performance, cost, and time. These can now be achieved. Our research, analysis, and decisions will support our goal of delighting our customer (Performance), while reducing our cost (Cost), and providing our product to our customers the moment they want it (Time). We must view this goal as a never-ending journey that continues to improve all three components because no one component is more important than the others. All three must be viewed with equal importance and balance.

SPC will provide the balance of knowledge that will allow us to achieve this quality objective. Now we are making regular and frequent ascensions to the summit of the Improvement Mountain to reach our goal of quality.

Problem Solutions

Problem Solution Page 21

For our earlier electricity data, to compute the average kilowatt consumption for the year 1980, individual consumption for each of the twelve months is added and then divided by the total number of months (twelve). The equation would read

$$\text{Total KWH consumption} = X_{Jan} + X_{Feb} + X_{Mar} + \ldots + X_{Dec}$$

$$= 2749 + 5405 + 3936 + \ldots + 3439$$

$$= 31387$$

The 1980 total KWH consumed is 31387 which is then divided by 12 to arrive at the average which is 2616 Kilowatt hours.

ELECTRICITY USAGE

	1980	**1981**	**1982**	**1983**
JANUARY	2749	4081	4982	3936
FEBRUARY	5405	5004	4299	5024
MARCH	3936	2708	3793	3239
APRIL	2232	2423	1870	
MAY	1079	1493	1514	
JUNE	1330	2197	1713	
JULY	2616	2918	1890	
AUGUST	2820	2041	1897	
SEPTEMBER	2321	2041	1923	
OCTOBER	1583	1494	1316	
NOVEMBER	1877	1859	1843	
DECEMBER	3439	3936	2919	
Average	2,616	2,683	2,497	4,066

The average for 1980, 1981, 1982, and 1983 is 2616, 2683, 2497, and 4066 respectively.

Problem Solution Page 54

From our analysis of the electricity usage, I have computed the maximum kilowatt usage for any one month based on only 1980, 1981, and 1982. My reasoning for not including 1983 is 1983 includes only three months of the highest usage. Using 1983 data will distort the reserve amount for higher than normal usage.

ELECTRICITY USAGE

	1980	**1981**	**1982**	**1983**
JANUARY	2749	4081	4982	3936
FEBRUARY	5405	5004	4299	5024
MARCH	3936	2708	3793	3239
APRIL	2232	2423	1870	
MAY	1079	1493	1514	
JUNE	1330	2197	1713	
JULY	2616	2918	1890	
AUGUST	2820	2041	1897	
SEPTEMBER	2321	2041	1923	
OCTOBER	1583	1494	1316	
NOVEMBER	1877	1859	1843	
DECEMBER	3439	3936	2919	
Average	2,616	2,683	2,497	4,066
Range	4,326	3,511	3,666	1,785
Std	1,211	1,112	1,211	900

Average of 1980,1981,1982: 2598
Standard Deviation: 1147

We are prepared to take a 5% risk of not having enough funds available to cover the usage. A 5% risk equates to a Z-value of 1.645.

$$X_i = X_{Ave} \pm z * Std$$

Now we can complete the equation to arrrive at the maximum planned budget KWH consumption.

$$X_i = X_{Ave} + z * Std$$

$$= 2598 + 1.645 * 1147$$

$$= 2598 + 1887$$

$$= 4485$$

This study will assume that the distribution of the kilowatts is normal, the cost of a kilowatt is $.10. Now the 4485KWH is multiplied times $.10 per KWH. This gives us an estimate of $448.50 that must be in reserve to cover our monthly electricity usage.

Problem Solution Page 151-153

Recap of the Completed Exercise
The table and charts on the next page show the correct answers for the exercise from the prior pages. The table below shows the original numbers for our problem.

1:00	2:00	3:00	4:00	5:00	6:00	7:00	8:00	9:00	10:00
50	55	60	45	40	50	55	60	45	40
51	56	61	46	41	51	56	61	46	41
52	57	62	47	42	52	57	62	47	42
49	54	59	44	39	49	54	59	44	39
48	53	58	43	38	48	53	58	43	38

The $\overline{X}$ control limits are 52.3 grams and 47.7 grams. The usual mistake in these calculations is the selection of the A_2 constant. The A_2 is .577 because the number of samples in each subgroup is five. The number of subgroups has nothing to do with the selection of the A_2 value.

The R control limit is 8.5 grams for the upper and zero for the lower because the variability cannot be negative. As for the $\overline{X}$, the typical mistake comes in the selection of the constants. We must remember that all constants are based on the number of samples in the subgroup which allows us to approximate three standard errors.

The actual individual readings are not used in the drawing of the chart but are only used in intermediate steps to calculate the data points and compute the control limits.

Note the R chart is flat because the control limits are based on part to part variation in the subgroup. Since the values are selected consecutively, it is no surprise that the limits are tight. The tight limits apply for both the $\overline{X}$ & R control charts since both sets are derived from the R values of maximum and minimum for the five consecutive samples.

We haven't discussed the interpretation of the control chart, but note the $\overline{X}$ chart is totally erratic when compared to the limits. Let's turn the page and investigate the subtle things that a control chart can monitor.

Problem Solution Page 151-153 (continued)

	1:00	2:00	3:00	4:00	5:00	6:00	7:00	8:00	9:00	10:00
$\bar{X}$	50	55	60	45	40	50	55	60	45	40
R	4	4	4	4	4	4	4	4	4	4

$\bar{\bar{X}} = 50.0$ $\quad \bar{R} = 4$

Control Limits for $\bar{x}$ *Chart*: UCL = 50.0 + (.577)(4) = 52.3

LCL = 50.0 - (.577)(4) = 47.7

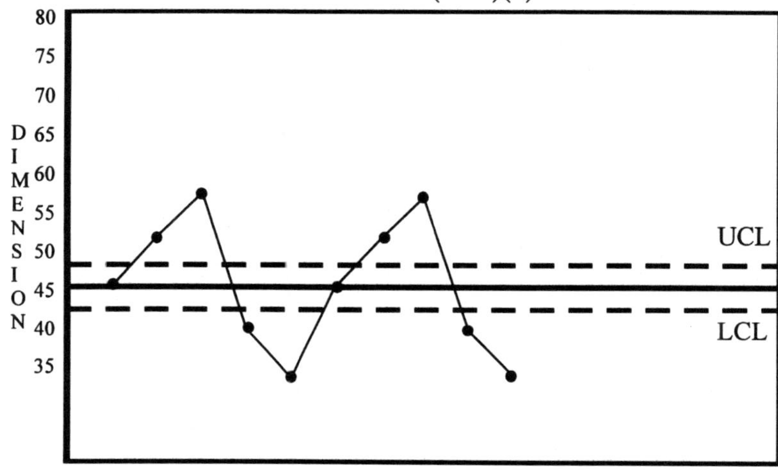

TIME OF MANUFACTURE: CHRONOLOGICAL ORDER

Control Limits for R Chart: (2.114)(4) = 8.5

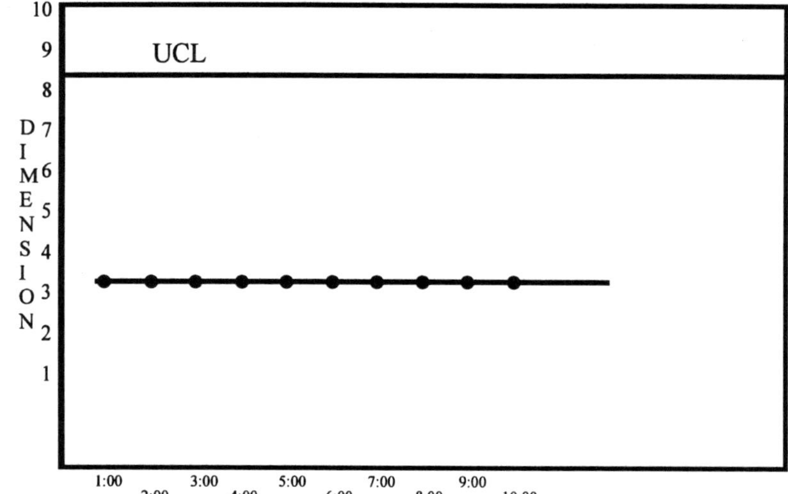

TIME OF MANUFACTURE: CHRONOLOGICAL ORDER

Problem Solution Page 268

A candy capability study of Chocolate Consistency

Let us now begin our capability study. The following readings are GooGoo chocolate consistency readings. Each reading is a test result from an individual GooGoo. Each row is the test results taken in a subgroup. Each subgroup is taken every 30 minutes.

TIME	OB1	OB2	OB3	OB4	OB5	X_BAR	Range
8:00	0.75655	0.76309	0.79555	0.85799	0.68696	0.77203	0.17103
8:30	0.73337	0.72810	0.83793	0.72416	0.68698	0.74211	0.15095
9:00	0.68130	0.68741	0.72701	0.64468	0.84080	0.71624	0.19612
9:30	0.60811	0.79303	0.90703	0.93568	0.75377	0.79952	0.32756
10:00	0.76692	0.62595	0.60721	0.69480	0.66086	0.67115	0.15971
10:30	0.73468	0.61227	0.66185	0.72843	0.55483	0.65841	0.17985
11:00	0.72128	0.75937	0.79861	0.81424	0.61291	0.74128	0.20133
11:30	0.76573	0.86283	0.66762	0.70669	0.61636	0.72385	0.24647
12:00	0.73689	0.80831	0.80004	0.69097	0.62378	0.73199	0.18453
12:30	0.67940	0.63179	0.64418	0.74433	0.70640	0.68122	0.11254
1:00	0.68170	0.77080	0.62278	0.57358	0.71586	0.67294	0.19722
1:30	0.84709	0.77807	0.59742	0.69553	0.81948	0.74752	0.24967
2:00	0.73915	0.80295	0.49851	0.81626	0.70454	0.71228	0.31775
2:30	0.87418	0.77712	0.92717	0.72820	0.76661	0.81466	0.19897
3:00	0.65501	0.79409	0.68381	0.76357	0.76406	0.73211	0.13908
3:30	0.77542	0.85875	0.54692	0.68720	0.71317	0.71629	0.31182
4:00	0.77416	0.72534	0.70543	0.64646	0.83704	0.73769	0.19058
4:30	0.60286	0.70500	0.68467	0.66985	0.67785	0.66805	0.10214
5:00	0.73883	0.67804	0.72443	0.68760	0.67617	0.70101	0.06267
5:30	0.78914	0.66123	0.62074	0.73161	0.86498	0.73354	0.24424
6:00	0.71017	0.77596	0.61749	0.65637	0.78831	0.70966	0.17082
6:30	0.70909	0.79797	0.72123	0.73636	0.77683	0.74830	0.08888
7:00	0.72207	0.57868	0.67210	0.79768	0.73411	0.70093	0.21900
7:30	0.67223	0.76367	0.65249	0.76464	0.67843	0.70629	0.11214
8:00	0.74534	0.77008	0.71271	0.62285	0.77221	0.72464	0.14936

The readings come from one chocolate machine with the chronological order of manufacture intact. Each time a reading is taken, five consecutive GooGoos are captured and tested. These five test points will be averaged to become our monitors on the control chart.

Control Chart Exercise

Using the data above, compute and prepare the data and control limits from the above chocolate consistency data. The completed control charts are found on page 269.

$\bar{\bar{X}}$ = .722548

$\bar{R}$ = .187377

Problem Solution Page 268 (continued)

$\bar{X}$ & R Chart Tables

Subgroup Size	A_2	d_2	D_3	D_4
2	1.880	1.128		3.267
3	1.023	1.693		2.574
4	0.729	2.059		2.282
5	0.577	2.326		2.114
6	0.483	2.534		2.004
7	0.419	2.704	0.076	1.924
8	0.373	2.847	0.136	1.864
9	0.337	2.970	0.184	1.816
10	0.308	3.078	0.223	1.777
11	0.285	3.173	0.256	1.744
12	0.266	3.258	0.283	1.717
13	0.249	3.336	0.307	1.693
14	0.235	3.407	0.328	1.672
15	0.223	3.472	0.347	1.653

$\bar{X}$ Chart

$$UCL_{AVE} = \bar{\bar{X}} + A_2\ \bar{R} \qquad LCL_{AVE} = \bar{\bar{X}} - A_2\ \bar{R}$$

Control Limits for $\bar{X}$ Chart:

UCL = .722548 + (.577)(.187377) = .830665

LCL = .722548 - (.577)(.187377) = .614432

R Chart

$$UCL_R = (D_4)\ (\bar{R}) \qquad LCL_R = (D_3)\ (\bar{R})$$

Control Limits for R Chart:

UCL_R = (2.114)(.187377) = .396115

(Limits shown on Page 269 were calculated by computer with A_2 & D_4 of more than three digits of precision and therefore vary slightly from the above answers.)

Problem Solution Page 292

Capability Exercise The GooGoo components' cost per gram is $.005. With this information, we can begin to analyze our strategic options. In our original proposal, the cost is based on a GooGoo that weighs 45 grams. By multiplying the weight of 45 grams times a component cost of $.005/gram, *the anticipated cost of a GooGoo is $.225*. For an hour's run, the cost of 3000 GooGoos would be 3000 GooGoos times $.225/GooGoo equaling $675.

The GooGoo facility is being run correctly and consistently as demonstrated by a control chart.

The following four situations are provided for us to investigate. Imagine that this is your candy company. We have computed the capability of each in the following table.

Situations	1	2	3	4
Is the Process In Control?	Yes	Yes	Yes	Yes
Average Weight	50gm	52gm	47gm	50gm
Standard Deviation of Weight	1.0gm	1.0gm	1.0gm	2.0gm
Distribution of Weight?	Normal	Normal	Normal	Normal
Z	5	7	2	2.5
% Defects	.0	.0	2.23	.62
CPK	1.67	2.33	.67	.833

Our strategic recommendation as to the best situation is very dependent on our unique situation based on the three components of quality. Remember quality is comprised of performance, cost, and time.

Appendix A: Glossary of Terms

Attribute Data	Data that is characterized as counts or percentages.
C Chart	Designed to detect change in Counts.
Capability Study	The strategic assessment of the facility's ability to make the product the customer wants.
Confidence Interval	The range of the confidence limits. This is a tactical tool for detecting change.
Confidence Limits	The tactical tool used to detect changes in average. This tool is the spot where the average will fall based on the risk that we have selected for our study.
Control Limits	A tactical tool for an operator to detect change in the usual running of a machine. Control limits are a special confidence limit that uses a Z equal to three.
CPK Index	The strategic performance indicator of the worst case specification. The bigger the CPK the better the process is functioning.
Discrete Data	See ***Attribute Data***.
Distribution	The shape of the spread of a set of data points.
Histogram	A bar chart that displays the shape of the data's distribution.

In Control	The condition when a process is consistent and stable in its natural state.
LCL_{AVE}	The **Lower Control Limit** is three (3) standard errors below the $\bar{\bar{X}}$ value.
LCL_R	The **Lower Control Limit** is three (3) standard errors below the $\bar{R}$ value.
Lower Specification Limits	The low limit where product is deemed defective.
Mean	The total of all data points divided by the number of data points. This is also referred to as Average, $\bar{X}$, $\bar{\bar{X}}$.
Median	The center value of a set of numbers in ascending sequence. Half the values of the set will be above with the remaining half below.
Out of Control	The situation when a change has been detected.
Outliers	Extremes or anomalies in the data. Outliers can be broken into the 2 groups of Observable Errors and Numeric Extremes.
P Chart	Designed to detect change in percents and proportions.
Pareto	A technique for providing focus to problem areas.
Population	**[Statistics]** Calculations based on measurement of every possible member in a group.

Prediction Interval — The strategic tool for showing the range of the prediction limits.

Prediction Limits — A strategic tool to assess the impact to the customer of the product that our machine is making. This tool is based on the use of individual readings.

R Chart — Designed to detect the movement or changes n the variability of the process.

Sample — **[Statistics]** Calculations based on measurement of all members in a subset of a group.

Specification Limit — The extremes that the customer will tolerate as good product.

Standard Deviation — *Of The Individuals (STD_{IND})*...the magnitude of the amount of spread between each individual value (variability).

Standard Deviation — *Of The Average (STD_{AVE})*...used in monitoring of patterns and trends. This value is referred to as Standard Error. This value is the amount of spread between averages.

Statistical Process Control — The identification that a machine has numerically demonstrated it is running correctly and consistently.

Statistical Quality Control — This is a set of numeric tools to monitor and maintain a characteristic of a process or product.

Strategic — The art of achieving a goal. This term is used in both decision-making and goal setting. For more information on this term additional reading can be found in the book *Become the Manager of the Future*.

Subgroup Sample results for a particular time frame.

Tactical The art of achieving a goal utilizing *available means*. This term is used in both decision-making and goal setting. For more information on this term additional reading can be found in the book *Become the Manager of the Future*.

Target The ultimate metric measurement that the customer desires in his product.

Tolerances The amount of incompetence that the customer will tolerate today. The tolerances are used to compute the specification limit.

U Chart A C chart that is offset to take care of varying units of measure.

UCL_{AVE} The **Upper Control Limit** is three (3) standard errors above the $\bar{\bar{X}}$ value.

UCL_R The **Upper Control Limit** is three (3) standard errors above the $\bar{R}$ value.

USL The designation for the upper specification limit.

Variable Data Data that is characterized by dimensional readings like length, weight, diameter, ,etc.

$\bar{\bar{X}}$ ***Chart*** Designed to detect the movement or change in the **Central Tendency** of the process.

Appendix B: Glossary of Constants & Symbols

$\boldsymbol{A_2}$ is the conversion factor to arrive at 3 standard errors for use in determining UCL & LCL of $\overline{X}$ chart.

$\boldsymbol{A_3}$ is the conversion factor to arrive at 3 standard errors for use in determining UCL & LSL of $\overline{X}$ and S Charts.

$\boldsymbol{B_3}$ is the conversion factor to arrive at 3 standard errors for use in determining LCL of S chart.

$\boldsymbol{B_4}$ is the conversion factor to arrive at 3 standard errors for use in determining UCL of S chart.

$\boldsymbol{C_4}$ is the conversion factor for determining Standard Deviation.

$\boldsymbol{d_2}$ is the conversion factor for determining Standard Deviation of the individuals.

$\boldsymbol{D_3}$ is the conversion factor to arrive at 3 standard errors for use in determining LCL of R chart.

$\boldsymbol{D_4}$ is the conversion factor to arrive at 3 standard errors for use in determining UCL of R chart.

$\boldsymbol{E_2}$ is the conversion factor to arrive at 3 standard errors for use in determining UCL & LSL of Individual Charts.

$\bar{R}$ is the benchmark of the **Central Tendency** of the process.

R is the range of each sub-group and becomes the statistic of interest.

$\bar{X}$ is the average of each sub-group and becomes the statistic of interest.

$\bar{\bar{X}}$ is the benchmark of the **Central Tendency** of the process.

Z the ratio of the distance of a value away from the average as it relates to the monitor of variability Standard Deviation.

Appendix C: Glossary of Formulas

$\overline{X}$ & R control chart

X_n is the data reading for the particular sample where n equals the sample.

$$\overline{X} = \frac{X_1 + X_2 + X_3 + \ldots + X_N}{N}$$

where N equals the number of samples in the sub-group and K equals the number of sub-groups.

$$R = \left| X_{Highest} - X_{Lowest} \right|$$

$$\overline{\overline{X}} = \frac{\overline{X}_1 + \overline{X}_2 + \overline{X}_3 + \ldots + \overline{X}_K}{K}$$

$$\overline{R} = \frac{R_1 + R_2 + R_3 + \ldots + R_K}{K}$$

$$UCL_{AVE} = \overline{\overline{X}} + A_2 \; \overline{R} \qquad LCL_{AVE} = \overline{\overline{X}} - A_2 \; \overline{R}$$

$$UCL_R = (D_4)(\overline{R}) \qquad LCL_R = (D_3)(\overline{R})$$

Capability Studies

$$Std_{Ind} = \sqrt{\frac{\sum_{i=1}^{N} (\overline{\overline{X}} - X_i)^2}{N - 1}} \quad \text{or} \quad Std_{Ind} = \frac{\overline{R}}{d_2}$$

$$\frac{STD_{Ind}}{\sqrt{n}} = STD_{Ave}$$

where n is the number of samples.

$$Z(USL) = \left| \frac{\text{Upper Spec Limit} - X_{Ave}}{STD_{IND}} \right| \qquad Z(LSL) = \left| \frac{\text{Lower Spec Limit} - X_{Ave}}{STD_{IND}} \right|$$

$$CPK = \frac{Minimum\ (Z_{USL}, Z_{LSL})}{3}$$

$\bar{X}$ and S Control Charts

$$s = \sqrt{\frac{\sum (\bar{X} - X_i)^2}{N - 1}}$$

$$\bar{S} = \frac{s_1 + s_2 + s_3 + \ldots + s_K}{K}$$

$\bar{X}$ Chart Control Limits

$$UCL_{AVE} = \bar{\bar{X}} + A_3 \ \bar{S}$$

$$LCL_{AVE} + \bar{\bar{X}} - A_3 \ \bar{S}$$

S Chart Control Limits

$$UCL_s = B_4 \ \bar{S}$$

$$LCL_s = B_3 \ \bar{S}$$

Individual Control Charts

$$R = | X_i - X_j |$$

$$\bar{R} = \frac{R_1 + R_2 + R_3 + \ldots + R_K}{K}$$ where K equals the number of sub-groups

$\bar{X}$ Chart Control Limits

$$UCL_{AVE} = \bar{X} + (E_2)(\bar{R})$$

$$LCL_{AVE} = \bar{X} - (E_2)(\bar{R})$$

Moving Range Control Limits

$$UCL_R = (D_4)(\bar{R})$$

$$LCL_R = (D_3) (\bar{R})$$

Appendix D: Control Chart Tables

$\bar{X}$ & R Chart Tables

Subgroup Size	A_2	d_2	D_3	D_4
2	1.880	1.128		3.267
3	1.023	1.693		2.574
4	0.729	2.059		2.282
5	0.577	2.326		2.114
6	0.483	2.534		2.004
7	0.419	2.704	0.076	1.924
8	0.373	2.847	0.136	1.864
9	0.337	2.970	0.184	1.816
10	0.308	3.078	0.223	1.777
11	0.285	3.173	0.256	1.744
12	0.266	3.258	0.283	1.717
13	0.249	3.336	0.307	1.693
14	0.235	3.407	0.328	1.672
15	0.223	3.472	0.347	1.653

$\bar{X}$ & S Chart Tables

Subgroup Size	A_3	C_4	B_3	B_4
10	0.975	0.9727	0.284	1.716
11	0.927	0.9754	0.321	1.679
12	0.886	0.9776	0.354	1.646
13	0.850	0.9794	0.382	1.618
14	0.817	0.9810	0.406	1.594
15	0.789	0.9823	0.428	1.572
16	.763	.9835	.448	1.552
17	.739	.9845	.466	1.534
18	.718	.9854	.482	1.518
19	.698	.9862	.497	1.503
20	.680	.9869	.510	1.490
21	.663	.9876	.523	1.477
22	.647	.9882	.534	1.466
23	.633	.9887	.545	1.455
24	.619	.9892	.555	1.445
25	.606	.9896	.565	1.435

Individual Chart Tables

Subgroup Freq	E_2	d_2	D_3	D_4
2	2.660	1.128	-	3.267
3	1.772	1.693	-	2.574
4	1.457	2.059	-	2.282
5	1.290	2.326	-	2.114
6	1.184	2.534	-	2.004
7	1.109	2.704	0.076	1.924
8	1.054	2.847	0.136	1.864
9	1.010	2.970	0.184	1.816

Appendix E: Normal Distribution Table

The table below allows you to look up the area (percent) of the Normal Distribution based on the ratio of the Z value. This table is a tail oriented probability table.

$$Z = \left| \frac{X_{Ave} - X_i}{STD} \right|$$

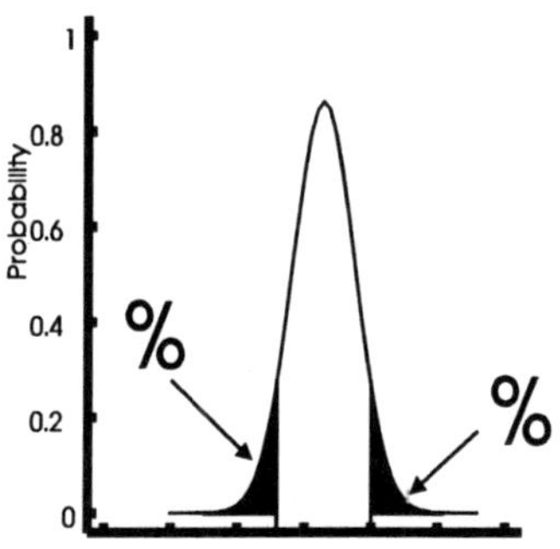

Z	X.X0	X.X1	X.X2	X.X3	X.X4	X.X5	X.X6	X.X7	X.X8	X.X9
4.5	0.000003	0.000003	0.000003	0.000003	0.000003	0.000003	0.000003	0.000002	0.000002	0.000002
4.0	0.000032	0.000030	0.000029	0.000028	0.000027	0.000026	0.000025	0.000024	0.000023	0.000022
3.5	0.000233	0.000224	0.000216	0.000208	0.000200	0.000193	0.000185	0.000179	0.000172	0.000165
3.0	0.001350	0.001306	0.001264	0.001223	0.001183	0.001144	0.001107	0.001070	0.001035	0.001001
2.9	0.001866	0.001807	0.001750	0.001695	0.001641	0.001589	0.001538	0.001489	0.001441	0.001395
2.8	0.002555	0.002477	0.002401	0.002327	0.002256	0.002186	0.002118	0.002052	0.001988	0.001926
2.7	0.003467	0.003364	0.003264	0.003167	0.003072	0.002980	0.002890	0.002803	0.002718	0.002635
2.6	0.004661	0.004527	0.004397	0.004269	0.004145	0.004025	0.003907	0.003793	0.003681	0.003573
2.5	0.006210	0.006037	0.005868	0.005703	0.005543	0.005386	0.005234	0.005085	0.004940	0.004799
2.4	0.008198	0.007976	0.007760	0.007549	0.007344	0.007143	0.006947	0.006756	0.006569	0.006387
2.3	0.010724	0.010444	0.010170	0.009903	0.009642	0.009387	0.009137	0.008894	0.008656	0.008424
2.2	0.013903	0.013553	0.013209	0.012874	0.012545	0.012224	0.011911	0.011604	0.011304	0.011011
2.1	0.017864	0.017429	0.017003	0.016586	0.016177	0.015778	0.015386	0.015003	0.014629	0.014262
2.0	0.022750	0.022216	0.021692	0.021178	0.020675	0.020182	0.019699	0.019226	0.018763	0.018309
1.9	0.028716	0.028067	0.027429	0.026803	0.026190	0.025588	0.024998	0.024419	0.023852	0.023295
1.8	0.035930	0.035148	0.034379	0.033625	0.032884	0.032157	0.031443	0.030742	0.030054	0.029379
1.7	0.044565	0.043633	0.042716	0.041815	0.040929	0.040059	0.039204	0.038364	0.037538	0.036727
1.6	0.054799	0.053699	0.052616	0.051551	0.050503	0.049471	0.048457	0.047460	0.046479	0.045514
1.5	0.066807	0.065522	0.064256	0.063008	0.061780	0.060571	0.059380	0.058208	0.057053	0.055917
1.4	0.080757	0.079270	0.077804	0.076359	0.074934	0.073529	0.072145	0.070781	0.069437	0.068112
1.3	0.096801	0.095098	0.093418	0.091759	0.090123	0.088508	0.086915	0.085344	0.083793	0.082264
1.2	0.115070	0.113140	0.111233	0.109349	0.107488	0.105650	0.103835	0.102042	0.100273	0.098525
1.1	0.135666	0.133500	0.131357	0.129238	0.127143	0.125072	0.123024	0.121001	0.119000	0.117023
1.0	0.158655	0.156248	0.153864	0.151505	0.149170	0.146859	0.144572	0.142310	0.140071	0.137857
0.9	0.184060	0.181411	0.178786	0.176186	0.173609	0.171056	0.168528	0.166023	0.163543	0.161087
0.8	0.211855	0.208970	0.206108	0.203269	0.200454	0.197662	0.194894	0.192150	0.189430	0.186733
0.7	0.241964	0.238852	0.235762	0.232695	0.229650	0.226627	0.223627	0.220650	0.217695	0.214764
0.6	0.274253	0.270931	0.267629	0.264347	0.261086	0.257846	0.254627	0.251429	0.248252	0.245097
0.5	0.308538	0.305026	0.301532	0.298056	0.294598	0.291160	0.287740	0.284339	0.280957	0.277595
0.4	0.344578	0.340903	0.337243	0.333598	0.329969	0.326355	0.322758	0.319178	0.315614	0.312067
0.3	0.382089	0.378281	0.374484	0.370700	0.366928	0.363169	0.359424	0.355691	0.351973	0.348268
0.2	0.420740	0.416834	0.412936	0.409046	0.405165	0.401294	0.397432	0.393580	0.389739	0.385908
0.1	0.460172	0.456205	0.452242	0.448283	0.444330	0.440382	0.436441	0.432505	0.428576	0.424655
0.0	0.500000	0.496011	0.492022	0.488033	0.484047	0.480061	0.476078	0.472097	0.468119	0.464144

Index

A

B

C

D

E

F

G

H

I

L

M

N

O

P

Q

R

S

T

U

V

W

X

Z

Notes